NATO Security through Science Series

This Series presents the results of scientific meetings supported under the NATO Programme for Security through Science (STS).

Meetings supported by the NATO STS Programme are in security-related priority areas of Defence Against Terrorism or Countering Other Threats to Security. The types of meeting supported are generally "Advanced Study Institutes" and "Advanced Research Workshops". The NATO STS Series collects together the results of these meetings. The meetings are co-organized by scientists from NATO countries and scientists from NATO's "Partner" or "Mediterranean Dialogue" countries. The observations and recommendations made at the meetings, as well as the contents of the volumes in the Series, reflect those of participants and contributors only; they should not necessarily be regarded as reflecting NATO views or policy.

Advanced Study Institutes (ASI) are high-level tutorial courses to convey the latest developments in a subject to an advanced-level audience

Advanced Research Workshops (ARW) are expert meetings where an intense but informal exchange of views at the frontiers of a subject aims at identifying directions for future action

Following a transformation of the programme in 2004 the Series has been re-named and re-organised. Recent volumes on topics not related to security, which result from meetings supported under the programme earlier, may be found in the NATO Science Series.

The Series is published by IOS Press, Amsterdam, and Springer, Dordrecht, in conjunction with the NATO Public Diplomacy Division.

Sub-Series

A. Chemistry and Biology	Springer
B. Physics and Biophysics	Springer
C. Environmental Security	Springer
D. Information and Communication Security	IOS Press
E. Human and Societal Dynamics	IOS Press

http://www.nato.int/science
http://www.springer.com
http://www.iospress.nl

Series C: Environmental Security – Vol. 8

Security of Water Supply Systems: From Source to Tap

Security of Water Supply Systems: From Source to Tap

edited by

Jaroslav Pollert
Czech Technical University,
Prague, Czech Republic

and

Bozidar Dedus
PRONING DHI d.o.o.,
Zagreb, Croatia

Springer

Published in cooperation with NATO Public Diplomacy Division

Proceedings of the NATO Advanced Research Workshop on
Security of Water Supply Systems: From Source to Tap
Murter, Croatia
27-31 May 2005

A C.I.P. Catalogue record for this book is available from the Library of Congress.

ISBN-13 1-4020-4562-X (HB)
ISBN-13 978-1-4020-4562-2 (HB)
ISBN-10 1-4020-4564-6 (e-book))
ISBN-13 978-1-4020-4564-6 (e-book)

Published by Springer,
P.O. Box 17, 3300 AA Dordrecht, The Netherlands.

www.springer.com

Printed on acid-free paper

Printed in the Netherlands.

TABLE OF CONTENTS

PREFACE

The provision of safe drinking water is one of the primary responsibilities of all governments, which address and share this responsibility through various levels of administration, ranging from the municipal to federal level, and further sharing of such responsibilities with public or private water utility companies. Recent reviews indicate the existence of significant vulnerabilities of all components of the infrastructure in general, among which the water supply systems are considered the most critical because of it's importance to human life. Indeed, such systems encompas huge number of stuctures, plants and devices that might become a target of sabotage, and they all may be found in major components of each water supply system: the raw water sources (usually a reservoir, a river intake, or groundwater aquifer), water purification plants (encompassing various treatment processes), and water distribution networks bringing potable water to the consumers.

Consequently, the reality of the post-September 11 situation forces the operators of water supply systems throught the world to examine the security and safety of their systems, it's vulnerability to intentional interference and sabotage with respect to quantity and quality of potable water. In assessing the system vulnerability, there is an urgent need to develop emergency response plans providing ways and means for alternative water supply for the moment of system operation disruption, and system remediation and recovery after the attack. Also where such plans already exist, there is a clear need to renew existing ones to sophisticated methods of possible attacks on water supply systems, where computer based activities are seen as possible way of provoking collapse of public water supply.

Based on mutual discussions among both ARW directors and their colleagues, the idea to apply for the ARW arose, and the Application was submitted in September 2004.

Holding the workshop in Croatia brings the focus to the region of Europe with many political challenges and needs for support by leading international bodies, such as NATO.

The idea to have ARW in Croatia was supported also by the fact, that there are still very "fresh" memories of the hostile war activities taking place in early 90's on Croatian soil when many water supply systems in Croatia were totally or partially damaged causing immense problems for people's lives (see the conclusions/information about the technical Tour to Zadar Water Supply Company) and huge economical damages for the national economy.

Thus, the proponents are convinced that this proposal addresses one of the highest priorities of the NATO security through science program in Europe, and at the critical time for action.

After receiving the NATO support grant, the organizers have confirmed keynote speakers, selected workshop participants, finalized the workshop programme, and held the workshop in Murter, Croatia. There were 42 full time participants at the workshop, from 12 countries, and additional observers from Croatia who also audited the workshop programme. Extensive experience of workshop participants in this field is reflected in the workshop proceedings, which comprise 16 selected papers.

The proceedings that follow reflect only the formal workshop presentations. Besides these papers, posters, and extensive discussions, there were many other ways of sharing and exchanging information among the participants. Hotel Colentum management in Murter, Croatia fulfilled all the necessary conditions which served a positive outcome of the conference. The venue itself was found by all participants very pleasant and comfortable, the atmosphere at the venue was really full of creative inventions and fruitful discussions. For this success, the editor and organizers are indebted to many who helped stage the workshop and produce its proceedings, as listed in the Acknowledgement.

Jaroslav Pollert
Prague, Czech Republic

Bozidar Dedus
Zagreb, Croatia

ACKNOWLEDGEMENT

This Advanced Research Workshop (ARW) resulted from hard work of many individuals and organizations. The workshop was proposed and directed by prof. Jaroslav Pollert, Czech Technical University in Prague, Department of Sanitary and Ecological Engineering, Czech Republic, and M.Sc. Božidar Deduš, Proning DHI d.o.o., Zagreb, Croatia. They were assisted by two other members of the workshop Organising Committee, Dr. Jiri Marsalek, National Water Research Institute, Environment Canada, Burlington, Canada, and Dr. Evzen Zeman DHI Hydroinform, Prague, Czech Republic.

The ARW was sponsored by NATO, Public Diplomacy Division, Collaborative Programmes Section, in the form of a grant. The preparatory work was conducted by the team from three institutons: Czech Technical University in Prague (led by Mrs. Rohanova), DHI Hydroinform Prague (led by Mrs. Nesvadbova), and Proning DHI d.o.o., Zagreb (led by Mrs. Milidrag). Local arrangements were carried out by Mr. Deduš, Mrs. Milidrag, Mrs. Nesvadbova, Dr. Zeman, prof. Pollert and Mrs. Rohanova. The organizers must give special thanks to companies: Crotel Gradnja d.o.o., Zagreb, Croatia; Hidroprojekt Consult d.o.o., Zagreb, Croatia, and Inženjerski Projektni Zavod d.d., Zagreb, Croatia, for their contribution to organization of the workshop. Special thanks are due to Dr. A.H. Jubier, Programme Director, Environmental Security, NATO, who provided liaison between the workshop organisers and NATO, and personally assisted with many tasks.

Also, special thanks go to Mr. Pjer Šimunović at the Ministry of Foreign Affairs of Croatia as Croatian National Coordinator for NATO for his help, assistance and suggestions in preparation of the workshop but as well for his attendance at the workshop.

Finally, the organisers are indebted to all the above contributors and, above all, to the participants, who made this workshop a memorable interactive learning experience for all involved.

LIST OF PARTICIPANTS

Directors

Pollert, J.	Czech Technical University, Dept. of Sanitary and Ecological Engineering Thakurova 7, 169 29 Praha 6 CZECH REPUBLIC
Dedus,B.	Proning DHI d.o.o.. Račkoga 3, 10 000 Zagreb, CROATIA

Key speakers

Cihakova, I.	Czech Technical University in Prague, Thakurova 7, Praha 6, CZECH REPUBLIC
Grayman, W.M.	W.M. Grayman Consulting Engineer, 321 Ritchie Ave., Cincinnati, USA
Ingeduld, P.	DHI Hydroinform a.s., Na Vrsich 5, Praha 10, CZECH REPUBLIC
Marsalek, J.	UWMP, 867 Lakeshore Rd., Burlington, CANADA
McBean, E.	School of Engng, University of Guelph, Ontario, CANADA
Samuels, W.B.	SAIC, 1410 Spring Hill Road, McLean, USA
Zeman, E.	DHI Hydroinform a.s., Na Vrsich 5, Praha 10, CZECH REPUBLIC

Other participants

Andročec, V.	Faculty of Civil Engineering Zagreb, Kačićeva 26, Zagreb, CROATIA
Banovec, P.	Faculty of Civil and Geodetic Engineering Ljubljana, Hajdrihova 28, Ljubljana, SLOVENIA
Berović, N.	Vodovod Zadar, Špire Brusine 17, Zadar, CROATIA
Brtko, B.	Slovak University of Technology, Radlinskeho 11, Bratislava, SLOVAK REPUBLIC
Ceausescu, M.D.	Hydraulics and Environmental Protection dept., Bvd. Lacul tei 124, Bucharest 38, ROMANIA
Čelica, T.	Ministarstvo za Okolje in Prostor, Dunajska cesta 48, Ljubljana, SLOVENIA
Gibala, T.	Slovak University of Technology, Radlinskeho 11, Bratislava, SLOVAK REPUBLIC
Havlik, V.	Czech Technical University, Thakurova 7, Praha 6, CZECH REPUBLIC
Horky, F.	Czech Technical University, Thakurova 7, Praha 6, CZECH REPUBLIC
Husarić, J.	Vodoopskrba i Odvodnja, Folnegovićeva 1, Zagreb, CROATIA
Kero, H.	Hidroprojekt Consult, Zagreb, CROATIA

Khoetsian, A. — Yerevan State University,
1 Alek Manoukian Street, Yerevan,
ARMENIA

Khomich, V. — Institute for Problems of Natural Resources Use & Ecology,
Staroborysovski tract 10, Minsk,
BELARUS

Knez, J. — Vodovod Zadar,
Špire Brusine 17, Zadar,
CROATIA

Kukharchyk, T. — Institute for Problems of Natural Resources Use & Ecology,
Staroborysovski tract 10, Minsk,
BELARUS

Loboda, A. — Ministarstvo za Okolje in Prostor,
Dunajska cesta 48, Ljubljana,
SLOVENIA

Mahrikova, I. — Slovak University of Technology,
Radlinskeho 11, Bratislava,
SLOVAK REPUBLIC

Meštrović, S. — Vodovod Zadar,
Špire Brusine 17, Zadar,
CROATIA

Morar, C. — Crisury Rivers Authority,
Ion Bogdan 35, Oradea,
ROMANIA

Perfler, R. — University of Natural Resources and Applied Life Sciences,
Muthgasse 18, Vienna,
AUSTRIA

Rohanova, B. — Czech Technical University,
Thakurova 7, Praha 6,
CZECH REPUBLIC

Rucka, J. — Brno University of Technology,
Zizkova 17, Brno,
CZECH REPUBLIC

Slavičkova, K.	Czech Technical University, Thakurova 7, Praha 6, CZECH REPUBLIC
Stancalie, G.	Romanian Meteorological Administration, 97 Soseaua Bucuresti - Ploiesti, sector 1, Bucharest, ROMANIA
Stanko, S.	Slovak University of Technology, Radlinskeho 11, Bratislava, SLOVAK REPUBLIC
Streng, O.	Crisury Rivers Authority, Ion Bogdan 35, Oradea, ROMANIA
Torosyan, G.	State Engineering University of Armenia, 105 Teryan, Yerevan, ARMENIA
Tothova, K.	Slovak University of Technology, Radlinskeho 11, Bratislava, SLOVAK REPUBLIC
Tuhovcak, L.	Brno University of Technology, Zizkova 17, Brno, CZECH REPUBLIC
Turkman, A.	Dokuz Eylul University, Buca Izmir, TURKEY
Uršić, M.	Faculty of Civil and Geodetic Engineering Ljubljana, Hajdrihova 28, Ljubljana, SLOVENIA
Van den Broeke, J.	Kiwa Water Research, Nieuwegein, NETHERLANDS
Vardanian, T.	Yerevan State University, 1 Alek Manoukian Street, Yerevan, ARMENIA

MANAGEMENT OF WATER SUPPLY SYSTEMS IN SLOVENIA AND ITS INTEGRATION ON THE STATE, REGIONAL AND LOCAL LEVEL

PRIMOŽ BANOVEC
*University of Ljubljana, Faculty of Civil and Geodetic Engineering, Jamova 2, 1000 Ljubljana, Slovenia**

Abstract. In this article, a new approach towards the management of water supply systems that is under implementation in the Republic of Slovenia will be presented. Current approaches are based on the integration of data at a horizontal level within the organization that provides the water supply service. One of the tasks of the service providers has always been reporting to the regulating authority. With increasing importance of the water supply service, the reporting requirements for other services is ever increasing and therefore a new approach is under way which aims at reporting of an extensive set of non-aggregated data on water supply systems. The new approach proves to be very efficient especially regarding the improved analytical capacity for other sectors and activities beside the regulatory agency itself, such as: regional spatial planning, emergency services, health, etc. The work has put forward the importance of good definition and management of spatial units where the service is provided – agglomerations. Water supply systems were defined on the basis of their hydraulic homogeneity, which enables back-tracing of water from the pipe to the source. The resulting spatial database has strong analytical capacity and is already under implementation as a useful decision-support tool.

Keywords: water supply system; management; vertical integration; data management; performance standards.

1. Introduction

The vital importance of water supply systems has always put them in a special focus regarding the legislation and institutional framework for determining their

---

**To whom correspondence should be addressed: University of Ljubljana, Faculty of Civil and Geodetic Engineering, Jamova 2, 1000 Ljubljana, Slovenia, pbanovec@fgg.uni-lj.si*

J. Pollert and B. Dedus (eds.), Security of Water Supply Systems: From Source to Tap, 1–7.

efficient and effective management. In Slovenia, water supply systems have been largely developed by 1980 covering approximately 95% of the population.

With the adoption of the strategy to associate with the European Union (EU), the focus has changed, and wastewater collection and treatment has gained in importance. This infrastructure represented for the biggest gap in public services related to the implementation of EU standards.

In recent years, water supply systems in Slovenia have again become a focus of attention, because of ecological and supply reliability issues, and also because of their lower priority in the past decade. Local communities, which are responsible for the water supply systems, are confronted with a set of demanding tasks in order to upgrade the entire water supply systems to face the challenges of the next decades.

1.1. FUNCTIONS RELATED TO THE EFFICIENT AND EFFECTIVE OPERATION OF WATER SUPPLY SYSTEMS

Basic functions of water supply systems are clearly to supply quality drinking water with sufficient pressure and discharge to satisfy the expectations of consumers. This requires integration of at least two sectors: the health sector with the responsibility for water quality, and the infrastructure sector with the responsibility for construction and O&M tasks.

1.2. ADDITIONAL RELATIONSHIPS AFFECTING THE MANAGMENET OF WATER SUPPLYS

Additional relationships can be defined corresponding to multi-perspective analysis (Linstone 1970). In the case of water supply, the following institutional aspects were defined:

- Integration within the water-cycle of the river basin – water management systems, according to the WFD[2], require development of tools to measure and model water abstractions and pollution loads to water bodies. To enable this, logical connections between water supply systems and water bodies from which they are supplied have to be established.
- Integration of water supply service into spatial planning procedures – water supply is one of the key services that enable population settlement and development of new production facilities.
- Emergency management – civil protection agencies perform emergency management in the case of natural and other disasters. Water supply systems provide the necessary water for basic human consumption and to satisfy fireflow demand, and also in case of emergencies.

When challenged with the task to develop a data structure that will enable horizontal and vertical integration of the data in order to enable decision support, it became clear that the common definition of service areas is of utmost importance.

2. Definition of agglomerations as a service areas

The definition of service areas for water supply systems is an important part of the process necessary for efficient development and management of water supply systems. The way they were defined depends to a large extent on the general availability of topographic and population data.

In the case of the Republic of Slovenia, the availability of different data on topography as well as data on population statistics in digital form is at a high level. The most important dataset describing population statistics is a registry of household numbers and population. The following approach was used to aggregate population data for the Republic of Slovenia. The entire area of Slovenia was divided into a regular polygon grid of 100 m x 100 m. For each polygon a query was performed in the registries regarding the number of inhabitants in the polygon. In the next step, adjacent non-empty polygons were dissolved into single agglomerations. The logical concept of agglomeration development is shown in figure 1.

Single non-void polygons were filtered out and defined agglomerations were communicated to the local communities for the verification. Different polygon size and distribution of polygons were tested and the resulting 100 x 100 m grid proved to be stable as well as of suitable resolution.

As a result of the procedure, 16,000 agglomerations for the entire Republic of Slovenia were defined as service areas, of which 3,266 have more than 50 inhabitants, comprising 82% of the population of Slovenia. The distribution of the population in the agglomerations can be seen in Figure 2.

The marginal value of 50 inhabitants was imposed by the implementation of the EU Drinking water directive. Due to the fact that the explicit limit value of 50 inhabitants could sometimes prove to be too restrictive, agglomerations with more than 40 inhabitants were considered for further analysis.

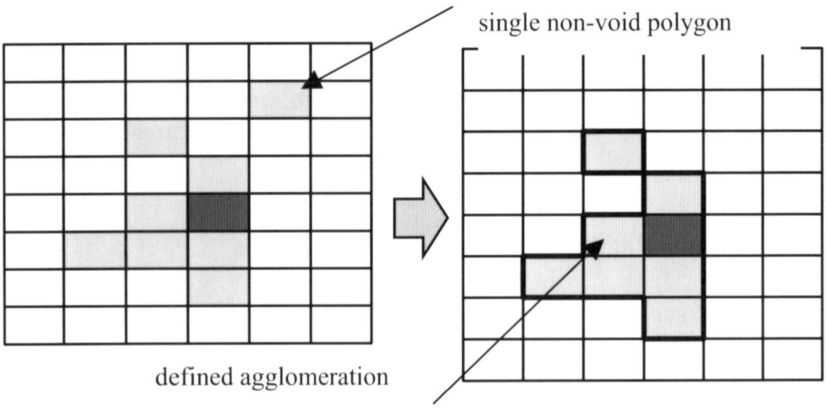

Figure 1. Procedure for the definition of the agglomerations

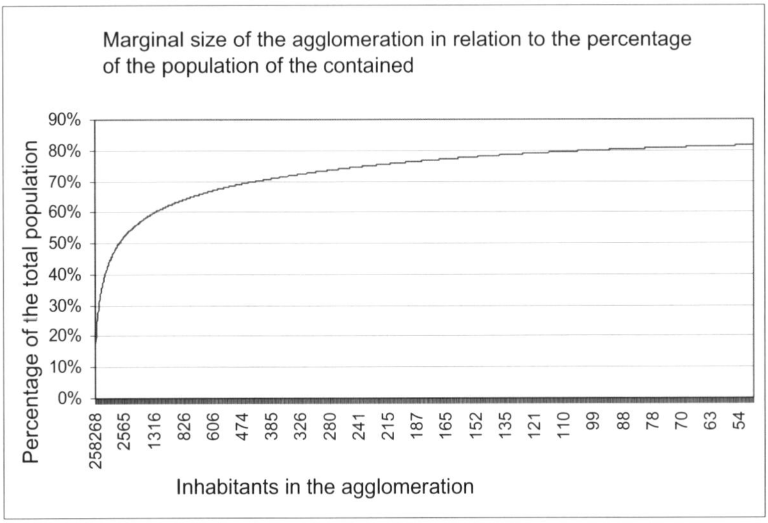

Figure 2. Marginal size of the agglomeration in relation to the percentage of the population contained

3. Definition of spatial entities of water supply systems (WSS)

Spatial entities for water supply systems are generally well known and broadly applied for different purposes, such as for operational management of WSS. Their use is essential for hydraulic modeling and derived analysis. In the case of vertical integration of data, another entity is of importance: the water supply system as whole. Certain data (i.e. water losses, pumping costs) could only be assigned to an overall water system. For this purpose a new definition of water supply system was developed:

> *Water supply system is a system of technical elements that supply water from its source to the users that is being managed by one legal entity (usually a public service provider) and operates mainly as an independent system hydraulically separated from other systems.*

By this definition, we have resolved a problem of transactions of water between two service providers (a clear legal, economical and technical border condition could be identified), as well as hydraulic consistency of individual WSS, by which we can also enable analysis of material flow (water) within one WSS. A key attribute for any water supply system is the ID of the public service provider which is simply its VAT code.

Beside the identification of a water supply system, the following key spatial entities were defined, that enabled vertical data integration:

- water resources - ID of all wells (groundwater), and surface water intakes; this ID serves as a link to the state service that manages groundwater bodies;
- reservoirs ID;
- pumping stations ID;
- drinking water facilities ID;

For all those entities, standard attribute tables were determined with which more detailed information about a specific entity could be collected and analyzed. The basic conceptual structure of the interactivity of the water supply systems is presented in Figure 3.

4. Development of the spatial data base of the WSS in Slovenia

In 2004 and 2005, reporting by more than 80 performers of public service in Slovenia and 194 local communities was used to establish the database of the water supply systems. The large number of public service providers is due to historical developments in Slovenia the reporting procedure itself was performed by electronic means, which required a detailed description of reporting procedures, organized education and individual technical support. Figure 4 shows the service areas in addition to the agglomerations that indicate the settlement patterns for Slovenia.

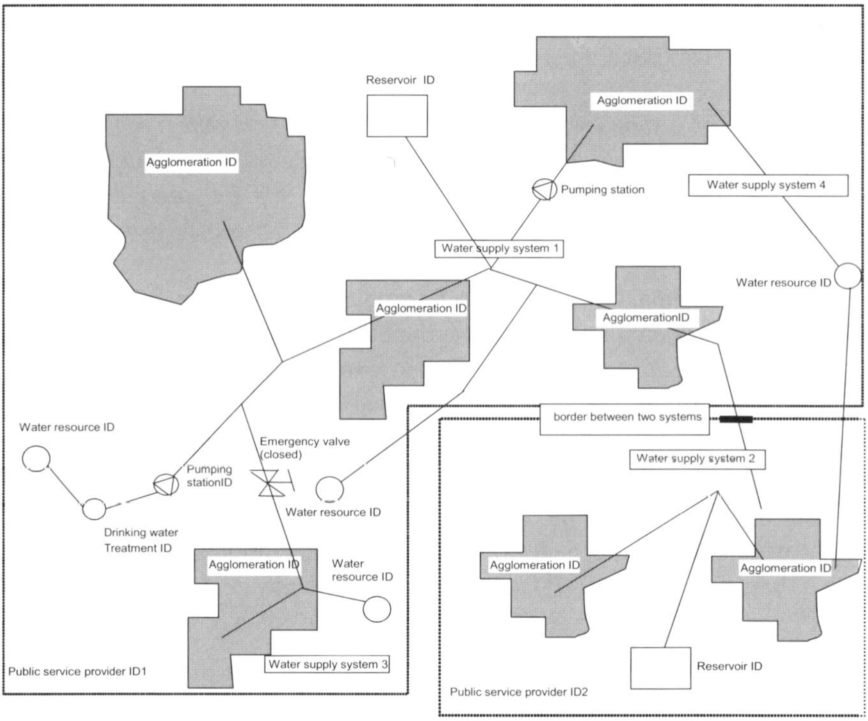

Figure 3. Conceptual scheme of different integration of different spatial entities

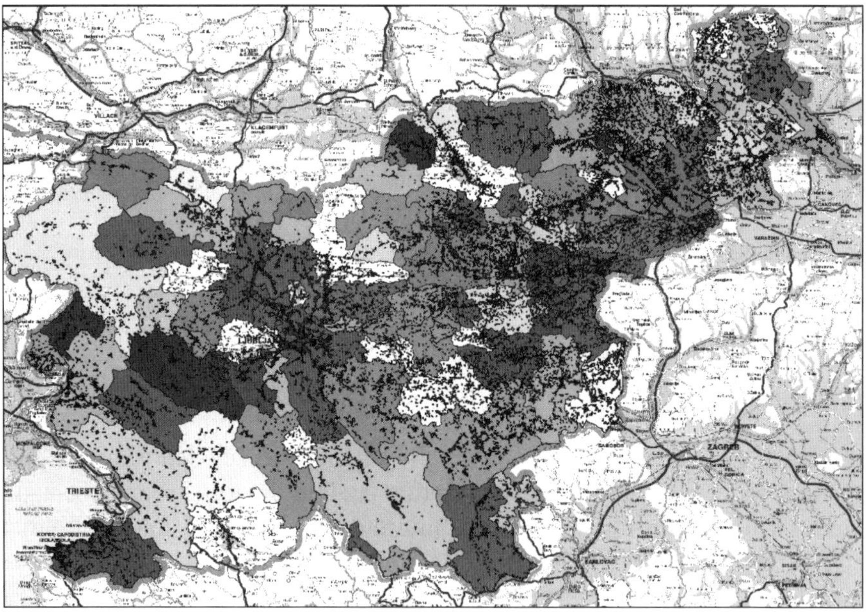

Figure 4. Overview of service providers and agglomerations in the RoS

The only spatial entity that was not part of the reporting at this stage was the spatial representation of water supply systems. Only attribute information was reported on water supply systems because they serve an important role in connecting all other entities (water resources, agglomerations, water treatment facilities, pumping stations, and reservoirs). The reporting scheme on the spatial representation of water supply systems (piping) is currently under preparation.

5. Conclusions

A comprehensive data model and reporting scheme was developed for the vertical integration of data on water supply systems. The developed data model is serving as a strategic support tool on the state and regional level, and is also of importance on local level.

Further improvement of the reporting system and database development is under way that will be able to perform partnership-based effective service among four main stakeholders in this area in Slovenia: local communities, state, water users, and providers of public services.

The implementation of this agglomeration-based approach will be demonstrated on some examples (modules), which are currently under development: (1) EU reporting module, (2) fire flow supply module, (3) water quality analysis module, (4) hydraulic performance module, and (5) water resources use module. Development of other modules is under consideration.

During the development of the reporting procedure and communication with different public entities, special attention was given to data protection and data availability as those high precision data which have extensive set of attributes have a dual character: (1) they represent public infrastructure and therefore should be made public, but (2) on the other hand they represent highly vulnerable infrastructure with potentially high risk impact to the public; therefore the access to these information should be limited.

Acknowledgement

The work on the reporting, database development and analysis was performed in the framework of contract No. 355-01-30/2004 assigned by the Ministry of Environment of the Republic of Slovenia.

FAILURE MECHANISMS AND MONITORING METHODOLOGIES PERTINENT FOR DETECTION OF EXPOSURE RISKS IN WATER DISTRIBUTION NETWORKS

EDWARD A. MCBEAN*
University of Guelph, Guelph, Ontario, Canada
(emcbean@uoguelph.ca)

Abstract. Source characteristics and failure mechanisms with potential to cause waterborne pathogen illness in the general population are described. Concerns with potential pathogen outbreaks are increasing in public awareness/attention as the occasions of incidence are increasing in occurrence and may be exacerbated by bioterrorism. Responsive efforts to ensure an effective and safe water supply system are increasing in complexity in part built upon increased understanding of pathogen fate and transport characteristics.

The potential for pathogen outbreaks to impact source water, and the resilience of the infrastructure including water treatment system and deterioration of the water distribution systems are considered. In response, efforts are described to improve monitoring capabilities, using indicator or surrogate parameters. An early warning system is described, representing an integrated comprehensive method for detecting, and assessing, a contamination event. The results of some experiments to allow 'trace back' as part of the investigations to determine where the contamination may have arisen are described.

Keywords: bioterrorism, surrogate monitoring, pathogens, disinfection byproducts, pipe breaks.

1. Introduction

Drinking water systems in developed countries involve delivery of a consumable product to communities, a product that is relied upon (trusted) by the population as fit for human consumption, and available upon demand. The importance of maintaining the existing extensive trust is critical, and relatively unique in terms of consumer service provided. However, for an array of

**To whom correspondence should be addressed:Edward McBean,University of Guelph, Guelph, Ontario, Canada (emcbean@uoguelph.ca)*

J. Pollert and B. Dedus (eds.), Security of Water Supply Systems: From Source to Tap, 9–17.

reasons, there are increasing concerns related to the ability to guarantee (i) the continuation of safe water, and (ii) water available upon demand.

Potential failure mechanisms to provide (i) and (ii) arise for reasons including, but not limited to, aging or poorly constructed infrastructure, extreme climatologic events, and intentional acts (e.g. malevolent and terrorists). Parts of the buried infrastructure in Canada are more than 100 years old in the central portions of major urban areas, and hence the infrastructure system is reaching its operating lifespan. Further, more severe climatologic events are stressing the integrity of the buried infrastructure systems, in part in response to identified climate change scenarios and causing interruptions to the water treatment effectiveness. The intent of terrorists is to instill fear or panic and undermine public confidence and/or to maximize injury potential (Haimes et al., 1998). These, and numerous other existing and/or potential failure mechanisms, all point to the need to improve the security of water supply systems.

The dimensions of the water supply system as defined herein include the water supply source, the water treatment system, and the water distribution system. The robustness of the water supply system in supplying safe water quality exists, in large part, as a result of the presence of multiple barriers (e.g. a succession of water treatment processes). If one barrier fails, then one or more of the alternative barriers provide(s) the necessary integrity to protect the quality of the supplied water to the population. However, while all of the preceding is very relevant to the entire water supply system, of the three specified elements of the water supply system, it is the water distribution system which is the most difficult for protecting the integrity of the water. Limited options are available to protect the consumer, once the water has been released to the distribution system. A failure that occurs downstream of the water treatment system bypasses the water system's primary defense systems.

This paper examines a number of the potential failure mechanisms within the water distribution system, and assesses the monitoring technologies available to warn of water quality problems within the distribution system.

2. Background

2.1. WATER DISTRIBUTION COMPONENT FAILURE

Widespread availability of clean water precipitated the most important change in human health which has occurred throughout all of modern history. Further, even that availability of water has had an enormous impact where, in response to the construction of four million tube wells in Bangladesh making water available to the rural population, the results significantly reduced infant mortality rate from 111 to 78 per 1000 live births between 1981 and 1996. However, subsequently identified exposures to arsenic and microbiological diseases from these tube wells have dramatically reduced this progress and hence the availability of water by itself, without adequate quality, is insufficient.

While the situation for Bangladesh has been dramatic, the number of waterborne disease outbreaks associated with drinking water by year and etiologic agent during the period from 1971 to 1998 has been substantial. The acknowledged number of waterborne disease outbreaks for individual years in the US is depicted in Figure 1, and total 691 in total. Further indications of the failure incidence arising in waterborne disease outbreaks are provided in Table 1.

Table 1. Examples of Disease Outbreaks

Location/Date	Pathogen	Result
New York, 06/96	*Salmonella*, poultry, manure	30 cases
U.K., 07/97	*E. coli*, cow manure	8 cases
Canada, Ontario 05-06-2000	*E.coli and Campyl*, cattle manure	1346 cases, 8 deaths
Canada, Sask. 03-05/2001	Animal or human waste	1907 cases
Developing countries/year	Polluted water (*Crypto, E. coli, giardia*, etc.)	1.7 million deaths

The causative features which resulted in the waterborne disease outbreaks listed in Figure 1 and Table 1 include dimensions of the quality of the water source, failure of the water treatment system, and ingress into the water distribution system. The various impacts have occurred on a frequent basis, even in the absence of terrorist acts and indicate the relatively high frequency of water supply systems failures. In drinking water distribution systems, there are few examples of real security improvements that have been developed and implemented during the last 60 years.

Indications of the array of failure mechanisms relevant to the distribution system are listed in Table 2.

Table 2. Examples of Water Distribution Component Failure

Examples of Mechanical Failure
- pipe breakage
- pump failure
- electrical power outage
- control valve failure

Hydraulic Failures
- old pipes with varying roughness, such as illustrated in Figure 2
- inadequate pipe size due to increased water demands
- insufficient in system storage capacity
Water Quality Failures
- concentrations of contaminants exceeding Maximum Concentration Level
- concentrations of residual disinfectant is below design level

The incidence of pipe failure contributes in a substantial way regarding the potential for water quality failures within the water distribution system. The causes of pipe failure include:

- changes in soil structure;
- temperature, moisture, chemical make-up, organic material and bacteria are common factors that can contribute to corrosion and eventually cause pipe failure;
- the soil environment, ground conditions weaken aging pipelines;
- drought periods followed by heavy rain can cause ground conditions to become unstable, causing corrosion-embrittled pipelines to fracture and break;
- temperature extremes, cold temperatures, frequently drive frost deeper into the ground, causing more rigid water pipes to break; and,
- failure to remedy the challenges caused by aggressive soil conditions.

An indication of the role of temperature within the Canadian context is provided by the cold temperatures which frequently drive damage deeper into the ground, causing more rigid water pipes to break. In fact, during the winter of 1995, Scarborough, a suburb of Toronto, experienced more than 160 breaks in one month alone. This corresponded to about one break for every four hours for 31 consecutive days.

2.2. POTENTIAL INCIDENCE OF PROBLEMS

Injection of chemicals or pathogens, or ingress of contaminated groundwater, downstream of the water supply and water treatment works, is the must vulnerable portion of the water system. Under the right conditions, it doesn't take much chemical (a few grains) to impact sporadically, a large community. There is no lack of chemicals which could be used (insecticides, solvents, corrosives, caustics). If added to the water supply system, the relative toxicities for some chemicals are listed in Table 3.

Table 3. Relative Toxicity for Some Chemicals

Chemical Agent	Acute Concentration (mg/L)
LSD	0.050
Cadmium	15
Hydrogen Cyanide	25
Arsenic	100-130
Dieldrin	5000

Fortunately, there are limited opportunities for entry of the chemicals to the water distribution system through injection of contaminants into the water distribution system (a pump to convert backpressure feeding via the typical residential service connection is able to only deliver restricted quantities. A tanker truck could deliver greater quantities via a fire hydrant but this would be much more noticeable and hence less likely).

Alternatively, if adequately mixed, one gram of feces mixed into 1 million gallons of water (3.785 million litres of water) could produce a 10 percent probability of rotavirus infection if 1 L of water was consumed (Rose, 2002). Unlike chemicals which would tend to have a dramatic and essentially immediate impact, pathogens would have a delayed and potentially more widespread effect. Further, the effectiveness of water treatment processes in removal of pathogens is less robust under some circumstances, and hence exposure scenarios to contaminated water are not restricted to failures via the water distribution system. Elevated levels of turbidity influence the effectiveness of removal of *Cryptospiridium* and *Giardia*, as illustrated in Figures 3 and 4. Given the relative ineffectiveness of chlorination as a destruction mechanism for *Cryptospiridium* and *Giardia*, there is significant potential for these as exposure scenarios unless UV is also within the water treatment sequence (the multiple barrier protection philosophy).

As a consequence of the delayed reaction to identify a pathogenic outbreak, and hence potentially more profound/widespread impact from pathogens, much of the attention herein is limited to microbiological features, not chemical exposure scenarios.

2.3. MONITORING OPPORTUNITIES THROUGH USE OF SURROGATES

Because the drinking water distribution system cannot be fully secured, there is need for monitoring the water quality within the distribution system. The use of indicator organisms by utilities and regulators to assure the microbial safety of drinking water is a widely applied and accepted practice. However, with improvements in epidemiological methods, it has become apparent that traditional coli form indicators are not always effective or efficient indicators of nonbacterial pathogens (Payment et al., 1985 and Rose et al., 1985). In addition, the presence of high numbers of indicator organisms which originate from animals which do not carry or transmit human infectious pathogens are of less concern for drinking water sources than contamination from humans or other carrier species. Complicating matters include the knowledgebase that pathogens of human health significance (e.g. *Ecoli* O157H7, *Salmonella* and *Campylobacter*) do not make the infected animal sick so the basis of identification of potential problems is not physically evident.

Alternatively, water supplies have been implicated in outbreaks of disease even though they were negative for coli form indicators (deNileon, 1998). Therefore, monitoring tools which allow for the delineation of specific microbial input sources are needed (Karlin et al., 1998).

Since there are innumerable pathogens, it is not possible to monitor them all. Most pathogen tests are not real time, but only provide information after the culture time. In the US and Canada, the traditional indicator organism system utilized for monitoring raw drinking water quality is enumeration of coli form bacteria using standardized EPA methods. Coli form bacteria are considered a reliable indicator of water quality and as a measure the adequacy of water treatment but recent research has demonstrated that coli forms do not provide absolute evidence of potential pathogen-bearing fecal contamination (Collins, et al., 1995). Indicator organism systems that provide greater specificity to pathogen presence have been researched. These systems include enumeration of *E. coli*, coli phages, fecal streptococci (FS), enterococci, and *spores* McFeters et al., 1978, Fujioka and Shizumura, 1985; Ashbolt et al., 1993; Toranzos and McFeters 1997).

Nevertheless, all of the above procedures provide test results which are 1 to 7 days old, allowing widespread consumption of the water before tests for indicator bacteria or viruses can be completed. Further, a high percentage of pathogens do not culture, requiring DNA procedures. The immediate tests for pathogens have, to date, proven to be expensive (e.g. as per Deininger, 2004).

In response, of interest are surrogates which could be used. Reliance upon surrogates represents a compromise between current scientific knowledge, need for conciseness, simplicity, and ease of utilization. The criteria for an ideal indicator or surrogate system include:

- be present or measurable when a pathogen is present and represent an imminent exposure risk;
- yield simple identifiers for a group of pathogens, without generating unacceptable levels of false positives;
- be sensitive to specific pathogen types and detectable; and,
- provide rapid identification of the present.

In response, the best opportunity at present to respond to the criteria for identification of pathogen excursions is to rely on water quality monitoring probes.

Currently, most municipalities in Canada only monitor within the water distribution system, chlorine residual. The adequacy of this approach, for the various reasons indicated above, is clear. Instead, a more effective approach is to rely upon a set of monitoring probes, including:

- pH where chemical changes via pK_a can be used to derive chemical concentration changes;
- conductivity could be used to provide an estimate of concentrations of dissolved solids;
- turbidity could be used as a possible indicator of suspended solids concentration changes;
- chlorine residual decreases could indicate presence of elevated organics and hence pathogens.

To avoid excessive false positives, it would be appropriate to use some excursion from typical fluctuations, before raising alarms. However, if the magnitude of the excursion is set too high, then false negatives would not be identified.

2.4. CHLORINATION

Chlorine residual is typically the only barrier that exists between discharge from the water treatment plant and the consumer tap. Hence, there are typical guidelines used to ensure protection of the water on basis of maintaining a residual chlorine level. Complicating attempts to protect the water supply by this mechanism include:

- chlorination is not effective for cryptospiridium and giardia. Hence, unless the water treatment processes include UV disinfection, there is no guarantee of water safety from these types of protozoa, both of which have very low infectivity levels (approximately 10 oocysts);
- extensive chlorination in water treatment is causative of significant potential for trihalogenmethane formation. The concern with formation of disinfection byproducts (DBPs) is very real with more than 10 percent of tested samples exceeding halo acetic acid guidelines of 0.06 mg/L (Zhu and McBean, 2004) and 9 percent of tested samples exceeding trihalogenmethanes guidelines of 0.08 mg/L (Zeng and McBean, 2005). It is noteworthy that more than 64% of samples in Ontario have DOC values greater than 2 mg/L which is the recommended level for reliance upon enhanced coagulation, as a means of minimizing DBP formation;

3. Modeling capabilities

While identification of contaminant excursions through monitoring is critical, the next stage is necessarily to identify the location of the source of contamination. Trace back algorithms are powerful in terms of identifying potential locations and involve utilization of algorithms which involve many applications of computer simulation/analytic models (e.g. Huang and McBean, 2005). As a simple demonstration, Figures 6 through 9 show the sequence of investigations for identifying spatial dimensions of a release from the Pump Station (at bottom of diagram) and how the spatial changes occur from 1 to 5 hours.

4. Conclusions

The need exists, and is increasing, for reliance upon multiple barriers to protect against exposure risks associated with water distribution networks. The combination of aging infrastructure, extreme meteorological events, and the potential for terrorist attacks, are all examples of potential excursions of contamination arising from consumption of water from water distribution systems. Increased knowledge of pathogens has indicated that reliance upon chlorination residuals for monitoring and fecal/total coli form as a surrogate of water quality is no longer adequate. The potential for excursions of contaminants within the water distribution system merit continuous, real-time analyses for timely detection of pathogens.

References

Ashbolt, N., Kueh, C., and Grohmann, G., 1993, Significance of Specific Bacterial Pathogens in the Assessment of Polluted Receiving Water of Sydney, Australia, Wat. Sci. Technology, 27, pp. 449-452.

Deininger, R., Personal Communication, February 2004.

deNileon, G., 1998, Water Supply Indicated in E. coli outbreak, Main Stream, 42, (8):1.

Fujioka, R., and Shizumura, L, 1985, Clostridium perfringens, a Reliable Indicator of Stream Water Quality, Journal Water Pollution Control Federation 57 (10), 986-992.

Karlin, R., Fayer, R., Arrowood, M., Noss, C., and Schoenen, D., 1998, Research Needs Panel Presented at the AWWA Source Water Protection Symposium: Afocus on waterborne pathogens, October 28-31, 1998, San Francisco, CA.

Haimes, Y.Y., et al., 1998, Reducing Vulnerability of Water Supply Systems to Attack", Journal of Infrastructure Systems, 4:164.

Huang, J., and McBean, E., "Comparison of Analytical and Simulation Approaches for Assessing Robustness of Reliability for Water Distribution Systems", Effective Modeling of Urban Water Systems, ed. W. James, K. Irvine, E. McBean, and R. Hicks, 2005.

LeChevallier, M., Norton, W., and Lee, R., 1991. "Giardia and Cryptosporidium spp. in Filtered Drinking Water Supplies", Applied and Environmental Microbiology, Sept, pp. 2617-2621.

McFeters, G., Shillinger, J., and Stuart, D., 1978, Alternative indicators of Water Contamination and Some Physiological Characteristics of Heterotrophic Bacteria in Water, In Evaluation of the Microbiology Standards for Drinking Water, C.W. Hendricks (ed.), Environmental Protection Agency.

Payment, P., Trudel, M., and Plante, R., 1985, Elimination of Virus and Indicator Bacteria at Each Step of Treatment during Preparation of Drinking Water at SEven Water Treatment Plants, Appl. Environ. Microbiol, 49, 1418-1428.

Rose, J., 2002, Water Quality Security, Envir. Sci and Technology, 36:247A.

Rose, J., Gerba, C., Singh, S., Toranzos, G., and Keswick, B., 1986, Isolation of Entero and Rotaviruses from a Drinking Water Facility, J. AWAA 78, pp. 56-61.

Toranzos, G., and McFeters, G., 1997, Detection of Indicator Microorganisms in Environmental Freshwaters and Drinking Waters, in Manual of Environmental Microbiology, Hurst et al., (eds.) ASM Press, Washington, D.C. pp. 184-194.

Zeng, W., and McBean, E., "Regression-based Modeling of Disinfection Byproduct Formation from Raw Water Source Treatment", in development.

Zhu, Z., McBean, E., and Zhou, H., "Optimization of Enhanced Coagulation in Water Treatment Using Bayesian Networks", Journal of Environmental Informatics, forthcoming 2005.

MONITORING OF ORGANIC MICRO CONTAMINANTS IN DRINKING WATER USING A SUBMERSIBLE UV/VIS SPECTROPHOTOMETER

JOEP VAN DEN BROEKE[1,*],
ALBERT BRANDT[1],
ANDREAS WEINGARTNER[2],
FRANZ HOFSTÄDTER[2]
[1]*Kiwa Water Research, the Netherlands;*
[2]*S::can Messtechnik GmbH, Austria*

Abstract. The UV-absorption of water at 254 nanometer has often been used as a measure for the presence of organic compounds (Conio et. al., 2002). Through analysis of the entire spectrum of UV and visible light, however, a lot of additional information can be obtained. Until recently, recording and analysis of the entire UV/Vis region was only possible using expensive equipment, suited only for operation in a well controlled laboratory environment. As the first on-line instruments for use in the field have now reached the market, it is time to determine the use and capabilities of these instruments.

Keywords: UV-Vis, Spectrophotometr, submersible, drinking water, monitoring

1. Introduction

For several years, portable, waterproof UV-probes have been on the market (Weingartner, 2003). These instruments can be used to record UV/Vis-spectra of liquids in-situ and in real-time. Using these "fingerprint"-spectra a number of specific parameters, such as turbidity, nitrate concentration and SAC_{254}, can be calculated (Langergraber et. al., 2003). This type of instruments is currently in use for monitoring influents of waste water treatment plants (Hofstaedter et. al., 2003; Langergraber et. al., 2003), monitoring of surface water quality (Langergraber et. al., 2004a) and for monitoring effluents of various industries

---

**To whom correspondence should be addressed: Joep van den Broeke, Kiwa Water Research, the Netherlands*

J. Pollert and B. Dedus (eds.), Security of Water Supply Systems: From Source to Tap, 19–29.

(Langergraber et. al., 2004b). In safeguarding drinking water quality and sources for drinking water production, however, there is relatively little experience.

An uninterrupted supply of drinking water of impeccable quality is the primary objective of the Dutch drinking water companies. Up-to-date knowledge on the composition of both raw- and drinking water is essential for achieving this goal. The employment of on-line monitoring systems allows the acquisition of water quality data and the UV-probe could be such a tool. Therefore Kiwa Water Research, within the Joints Research Programme executed on behalf of the Dutch drinking water companies, has evaluated such a UV-probe for applications in the drinking water supply.

2. The instrument

For the evaluation, the Spectro::lyscr (Figure 1) of the Austrian company S::can Messtechnik was selected, as this probe is one of the most widely used of its kind. This Spectro::lyser is a spectrophotometer that records the absorption spectrum of a water sample in the 200 – 735 nm wavelength range. The resulting fingerprint spectrum is used for the characterization of the sampled water (Langergraber et. al., 2003). Furthermore, the fingerprint is used to monitor changes in the water composition, and offers the possibilities to set alarm levels on the basis of the magnitude of the variations in the spectrum. A measurement cycle takes 45 seconds, making possible a high measuring frequency and rapid detection of changes. The fingerprint is also used to derive more specific parameters, such as turbidity, nitrate concentration, SAC_{254}, TOC and DOC.

The instrument uses a "dual-beam" configuration, in which one light beam passes through the sample while a second is used as internal reference. This ensures stability of the measurement over extended periods of time. The sensitivity of the probe for UV-absorbing organic components in the water sample is determined by the path length of the measuring compartment. Instruments with a number of path lengths are available, and for this project the instrument with the longest path length (100 mm) was selected as this offers the highest sensitivity. For use in applications with matrices with a high load of dissolved organics and/or high loads of suspended solids a less sensitive instrument, with a shorter path length, is required.

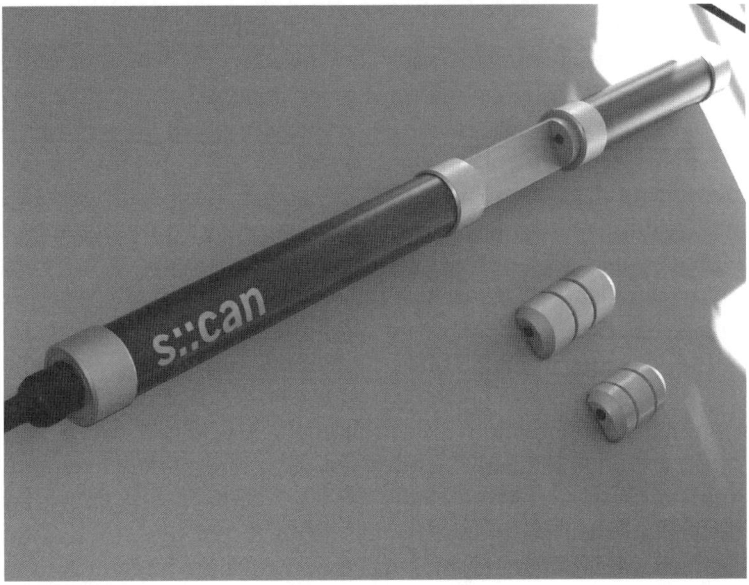

Figure 1. The evaluated instrument, together with inserts (right) for shortening the path length

3. Approach

The UV-probe was tested under a number of operational conditions: both in the laboratory and in the field the performance of the instrument in detection of contaminants in (drinking) water was evaluated. The results obtained with the UV-probe were compared with the data obtained with a laboratory spectrophotometer and with turbidity meters operated for routine measurements.

4. Results and discussion

4. 1. OFF-LINE USE OF THE UV-PROBE

To verify the accuracy of the UV-probe a number of short experiments was done in which the instrument was used as a conventional, off-line, UV spectrophotometer. Drinking water samples were analyzed using both the probe and a laboratory spectrophotometer, and identical results were obtained. Subsequently, a calibration curve was made for the parameter nitrate. This showed a linear response of the UV-probe for nitrate within the concentration range indicated by the manufacturer. The concentrations calculated by the software, however, did not correspond with the actual nitrate concentrations in the samples. This indicates that the software calibration (Langergraber et. al., 2003) provided with the instrument is not suited for quantitative analyses. This 'global calibration' is a software application that uses the average response of hundreds of measurements and is suitable for an approximation of concentrations in a particular matrix. For accurate measurements, however, it is not reliable and a local calibration for the water type at a measuring location is required. This requirement is valid for all specific parameters that can be determined with the UV-probe.

The capability of the instrument to detect and identify organic compounds in drinking water was evaluated by analyzing spiking water samples with specific contaminants. This part of the study focused on the detection of toxic compounds, such as pesticides and (stimulants for) chemical warfare agents (Figure 2) (Steiner et. al., 2003). Drinking water samples, spiked with a selection of compounds that are considered as most likely for use in the event of an intentional contamination of a drinking water supply, were analyzed. For each of the used contaminants calibration curves were obtained, and the lower detection limits were derived (table 1). This showed that compounds with a high UV-absorbance could be detected down to the μg/L concentration level. As there is no (chromatographic) separation of the compounds present in the water samples, a superposition of the absorption corresponding to the added compounds and that corresponding to the matrix is recorded. As a result, it was only possible to detect changes in water composition, and not to identify the spiked compounds on the basis of their UV-spectrum.

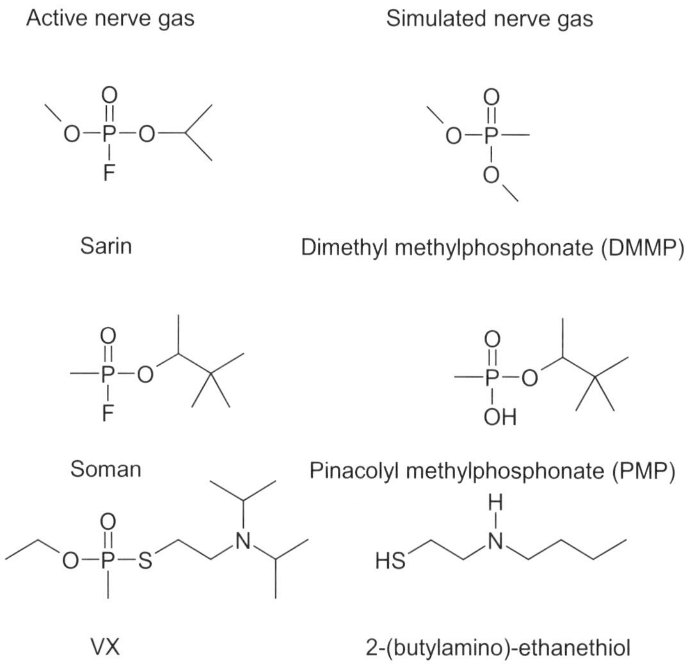

Figure 2. Nerve gas simulates measured using the Spectro::lyser

The capability of the instrument to detect and identify organic compounds in drinking water was evaluated by analyzing spiking water samples with specific contaminants. This part of the study focused on the detection of toxic compounds, such as pesticides and (simulates for) chemical warfare agents (Figure 2). Drinking water samples, spiked with a selection of compounds that are considered as most likely for use in the event of an intentional contamination of a drinking water supply, were analyzed. For each of the used contaminants calibration curves were obtained. This showed that compounds with a high UV-absorbance could be detected down to the μg/L concentration level (Table 1). As these is no (chromatographic) separation of compounds in the water, a superposition of the absorption corresponding to the added compounds and that corresponding to the matrix is recorded. As a result it was only possible to detect changes in water composition, and not possible to identify the spiked compounds on the basis of their UV-spectrum.

The capability of UV-spectroscopy to measure differences in water composition, and thus changes in water quality, was illustrated by the analysis of 3 drinking water samples. The recorded spectra of tap water from 3 geographically closely related locations, which are all served by different water treatment plants using ground water as raw water source, show distinctive differences. These differences (Figure 3) are mainly the result of different concentrations of humic acids in the water.

Table 1. Detection limits for specific organic compounds in drinking water

Compounds	Detection limit (mg/L)
PMP	50
DMMP	50
Azinphos-methyl	10
Methamidophos	1
Oxamyl	0.1
Mevinphos	0.1
Aldicarb	0.1
Benzene	0.05
Isoproturon	0.05
Linuron	0.001

[a]: 2-(Butylamino)-ethanethiol was also measured but no detection limit was established as the compound contained an unidentified impurity with a very high UV-absorption coefficient.

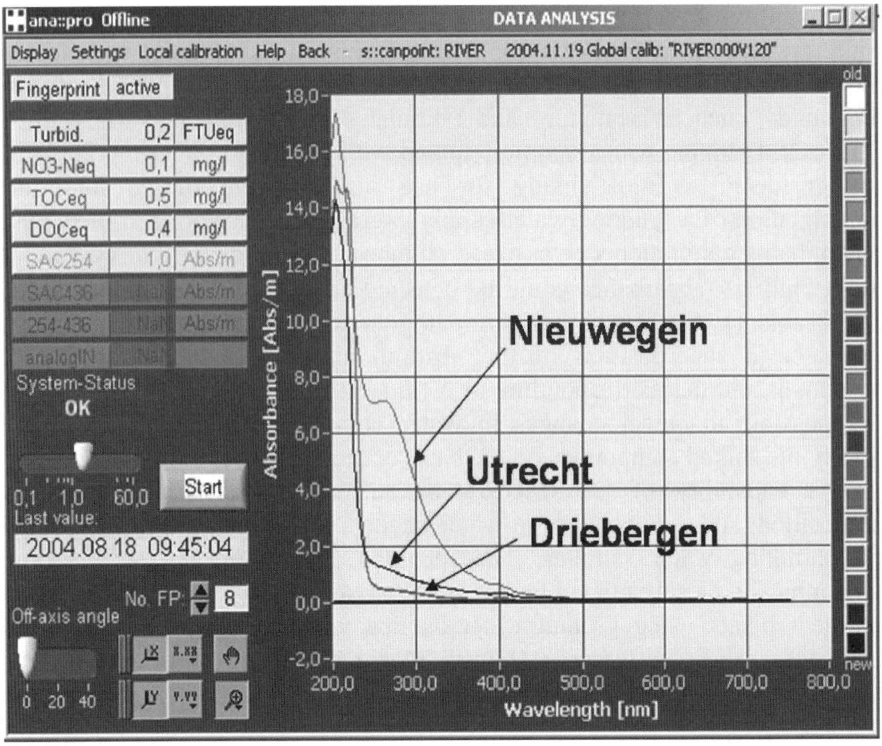

Figure 3. Fingerprint of drinking water obtained at three closely related locations

4.2. ON-LINE USE OF THE UV-PROBE

After the off-line experiments, the UV-probe was evaluated as on-line sensor in three different set-ups: on-line monitoring of drinking water (Nieuwegein, The Netherlands), on-line monitoring of surface water used for drinking water production (Lekkanaa, Nieuwegein, The Netherlands) and on-line monitoring of pre-treated water (at Waterwinstation Cornelis Biemond, Nieuwegein, The Netherlands). While monitoring drinking water the fluctuations in the measured parameters due to random noise were found to be insignificant. To verify this truly reflected water composition and was not caused by insensitivity of the probe, and also to verify the reproducibility of the signal, the drinking water was spiked with ascorbic acid (vitamin C) and benzoic acid. Benzoic acid was added so that a concentration in the drinking water of 5 mg/L was reached and with pulses during 15 minutes, repeated every 2 hours. Vitamin C was added in similar fashion, but each addition was performed 1 hour after the addition of benzoic acid. The concentration of vitamin C was gradually reduced to 0.2 mg/L, which was the minimum concentration that could be dosed using the set-up applied. All these additions were clearly observed (Figure 4); at a vitamin C concentration of 0.2 mg/L the standard deviation in the measured absorption during 10 subsequent was below 1%.

Identical results were obtained when monitoring pre-treated water; slow fluctuations in the matrix were observed, but no high peak concentrations or rapid changes were detected. The results of the turbidity measurements were compared with these of a turbidity meter installed at the same location. This showed that the detected trends in turbidity were comparable for the two instruments, however, the absolute values produced by the UV-probe differed significantly (> 500%) from these of the turbidity meter. As the turbidity meter was operated and calibrated using validated procedures, this indicates that the values produced by the UV-probe were not correct. Local calibration of the UV-probe was required to obtain quantitative readings on turbidity in this matrix. During the monitoring of the pre-treated water, vitamin C was dosed using the set-up as used for the experiments in drinking water. This showed the same sensitivity and reproducibility observed while spiking drinking water.

While monitoring surface water, containing a much higher sediment load than the matrices sampled so far, the path length of 100 mm was found to be unsuitable. A probe with a 10 mm path length was required to perform measurements. Within several days, however, it was found that deposition of silt on the probe still hampered the measurements (figure 5). This could have been solved using an automatic cleaning system, which cleans the optical windows using pressurized air. As this system was not installed on the evaluated instrument, however, the optical windows were clean manually at least once a week.

During the monitoring of the surface water, strong fluctuations in turbidity, TOC, DOC and UV-absorption at 254 nanometer were observed, while the measured nitrate concentration remained stable. These observed fluctuations in turbidity were identical to those observed with the turbidity meter installed at this location. In this case the absolute values for turbidity obtained with both instruments were similar (differences in the order of 10 – 50 % were observed); indicating that for this matrix the global calibration was a good approximation. The large fluctuations in the water composition were caused by frequent shipping and construction activities on the canal. This explains why all parameters related to suspended particles fluctuate strongly, whereas the nitrate concentration does not. The latter is less influenced by stirring up of silt from the bottom of the canal than the other parameters.

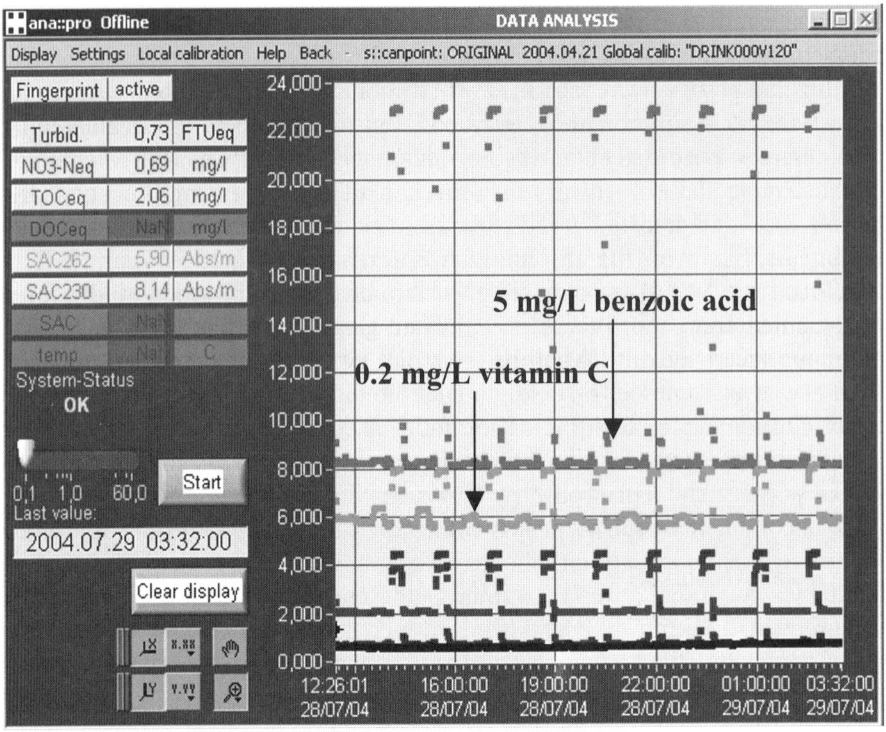

Figure 4. Monitoring of drinking water spiked with vitamin C (0.2 mg/L) en benzoic acid (5 mg/L). Besides TOC (red) the extinction at the wavelengths of maximum absorption for benzoic acid (230 nm, purple) and vitamin C (262 nm, orange) were determined

Despite the rapid fluctuations in water composition, it was still possible to detect a brief contamination event. Using a special alarm algorithm, and after training of the system to ensure it will not respond to natural fluctuations (Langergraber 2004a), this software tool was able to distinguish a brief spike corresponding to an organic compound. Using the alarm tool, this event could clearly be detected, while it was almost impossible to discern this event using the fingerprint spectra. The nature of the contaminant could not be confirmed using the UV-probe, but from routine sampling and monitoring at the Lekkanaal location it is known that at the time of the alarm, a high concentration of ftalates was present in the water.

Figure 5. Deposition of sediments can hamper measurement. Cleaning, e.g. using an automated cleaning system is required for proper operations in water with a high fouling potential

5. Conclusions

This evaluation has shown that the UV-probe is an instrument that can be used both in the laboratory and in the field. It can be used to detect (specific) contaminants in water, sometimes down to the µg/L level, but it is not capable of identifying such contaminants. For monitoring and detecting (natural) fluctuations in water quality the UV-probe proved to be an effective tool. In this type of application, the UV-probe provides a lot of additional information in comparison with conventional sensors, such as turbidity meters and single wavelength UV-detectors. The added value of UV-spectroscopy over such systems can be applied for security applications and in process control, where characterization of water is highly valuable.

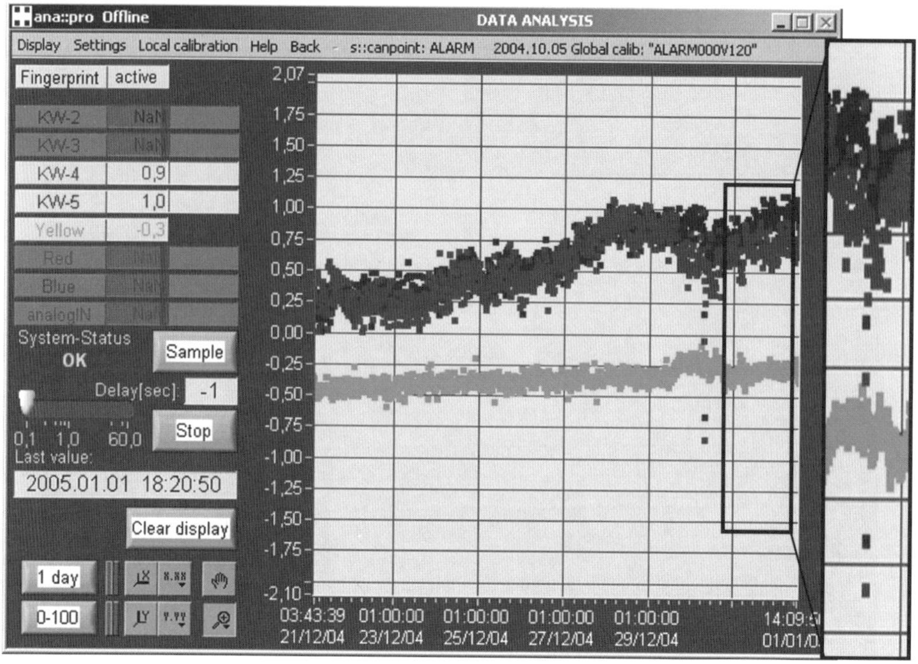

Figure 6. Detection of a water quality deviation using the alarm software

Acknowledgements

This work was funded by and executed within the Joint Research Programme of the Dutch drinking water companies (BTO project 111508.030).

References

Conio, O.; Chioetto, M.; Hargesheimer, E. (2002): Organic Monitors. In Hagesheimer, E.; Conio, O.; Popovicova, J. (Eds.): On-line monitoring for Drinking Water Utilities, Awwa Research Foundation, USA, 163 – 202.

Hofstaedter, F.; Ertl, T.; Langergraber, G.; Lettl, W.; Weingartner, A. (2003): On-line nitrate monitoring in sewers using UV/Vis spectroscopy. In: Wanner, J.; Sykora, V. (Eds.): proceedings of the 5th international conference of ACE CR "Odpadni vody – Wasterwater 2003", 13 – 15 May 2003, Olomouc, Czech Republic, 341 – 344.

Langergraber, G.; Fleischmann, N.; Hofstaedter, F. (2003): A multivariate calibration procedure for UV/Vis spectrometric quantification of organic matter and nitrate in wastewater. *Wat. Sci. Tech.* **47**(2), 63 – 71.

Langergraber, G.; Weingartner, A.; Fleischmann, N. (2004a): Time- resolved delta spectrometry: A method to define alarm parameters from spectral data. In Langergraber, G.; Winkler, S; Fleischmann, N.; Pressl, A; Haberl, R. (Eds.): proceedings of the 2nd international IWA conference on Automation in Water Quality Monitoring, 19 – 20 April, Vienna, Austria, 93– 100.

Langergraber, G.; Fleischmann, N.; Hofstaedter, F.; Weingartner, A. (2004b): Monitoring of a paper mill wastewater treatment plan using UV/Vis spectroscopy. *Wat. Sci. Tech.* **49**(1), 9 – 14.

Steiner, W. E.; Clowers, B. H.; Haigh, P. E.; Hill, H. H. (2003): Secondary Ionisation of Chemical Warfare Agent Simulants: Atmospheric Pressure Ion Mobility Time-of-Flight Mass Spectrometry. *Anal. Chem.* **75**, 6068 – 6076.

Weingartner, A. (2003): New Applications of s::can's ultra compact submersible UV-vis-spectrometers. *Int. Env. Tech.* **13**(2), 146.

EXPECTED DEVELOPMENT IN THE SUPPLY AND DISTRIBUTION OF DRINKING WATER IN CZECH REPUBLIC

IVA CIHAKOVA*
Czech Technical University in Prague, Department of Sanitary and Ecological Engineering, Prague, Czech republic,

Abstract. The information on the anticipated development of the supply and distribution of drinking water is important especially considering the development planning of regions, the decision-making about and design of the water supply systems, their capacity and size, and also their protection against major events, which include e.g. vandalism and terrorism. The question of lowering the losses in water distribution networks and their solution is very close to the prevention against these events.

Keywords: trends in water consumption, regulation in water consumption, price of water, water losses, jeopardy of water distribution system

1. Introduction

Water consumption development in principle shows how advanced a country is in terms of culture and technology, with all the positive and negative aspects. When looking at characteristic data, which describes the total volume of water consumed, while considering the total population, gross domestic product, environmental trends and drinking water availability issues, the views can be quite varied. Nonetheless, the primary aspect is that we deal with managing a limited natural resource. When looking at drinking water management globally and considering the recent political trends, we can discern two significant periods of the recent past. These can be described with IWSA programs implemented in 1970-1980 and 1980-1990. The first one was called "Drinking Water to All", while during the second period there was a clear shift toward accepting economic requirements under the title "Drinking Water is a Product". Presently, this attribute needs to be amended since we can definitely declare

To whom correspondence should be addressed: Iva Cihakova ,Czech Technical University in Prague, Department of Sanitary and Ecological Engineering,Prague, Czech republic, cihakova@fsv.cvut.cz

J. Pollert and B. Dedus (eds.), Security of Water Supply Systems: From Source to Tap, 31–38.

that "Drinking Water is Our Strategic Commodity", and its protection combined with protecting the water supply system will cost us considerable funds.

The water supply system is always designed for water consumption during a given period of time, providing that it is possible to estimate and predict crucial factors such as development of the population, purchasing power, technical and technology levels as well as political situation in the given region; as such, it is negatively affected by contingent (unpredictable) events. Water management companies therefore strive to learn about methods how to regulate water consumption. Water consumption is likely to be reduced significantly during major negative events, e.g. environmental incidents. Reducing water consumption, however positive it may be in terms of maintaining the quantity of quality drinking water, is not a positive trend for water management companies, and it often leads to pressures on increasing the water price. Studies have been published on water demand regulation in line with the adjusted price and price plans. Variable financing makes it possible to set the water price more sensitively to make it acceptable, available, while ensuring that higher prices can be charged for high water consumption. It is evidently quite difficult to establish an ideal balance in the price of this indispensable resource for people. In a number of countries, drinking water prices are at least partially regulated by the government, and water prices are often subsidized by municipal and regional administration bodies. In the Czech Republic, water prices are still regulated by the government. However, better financial performance has been recorded by organizations in which the city or region exercises a certain influence, or which are subsidized by cities and regions.

2. Expected development – saving water and improving the supplied water quality

- Trend and forecast in the development of requirements for water
- Improving the water distribution systemSuitable water mains diagnostics
- Technical development
- Funding methods When monitoring water consumption development,

we analyzed data from the previous years and compared actual volumes of water consumed in different locations. One of the descriptive values is for example water consumption in households, expressed in liters per person per day.

Water consumption clearly declined in a number of cities also after 1990. In Prague, this decline can primarily be attributed to economic factors (significant water price increase). It also reflects the overall social consequences after 1989 when the governmental economic policy changed, and so did the strategic goals and priorities. Cities with significantly stable financial standing show declining water consumption (Hamburg, Nürnberg, Copenhagen); in contrast, we can see alternating trends in cities such as Vienna and Frankfurt am Main. Berlin, a city reunited after the long years of the Cold War, clearly shows a dynamic decline in the first half of the 1990s, caused mainly by the situation in the former Eastern Zone. In the years that followed, water consumption began to grow again. Governmental institutions have gradually been relocated to the city, which induced other organizations and businesses to establish there offices or to relocate their headquarters to the capital.

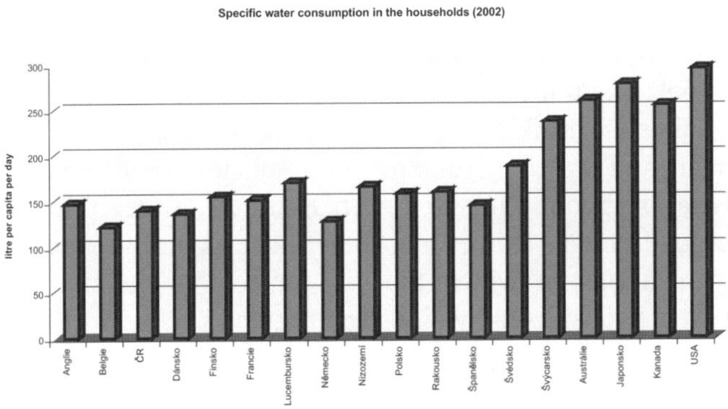

Figure1. Specific water consumption in the households

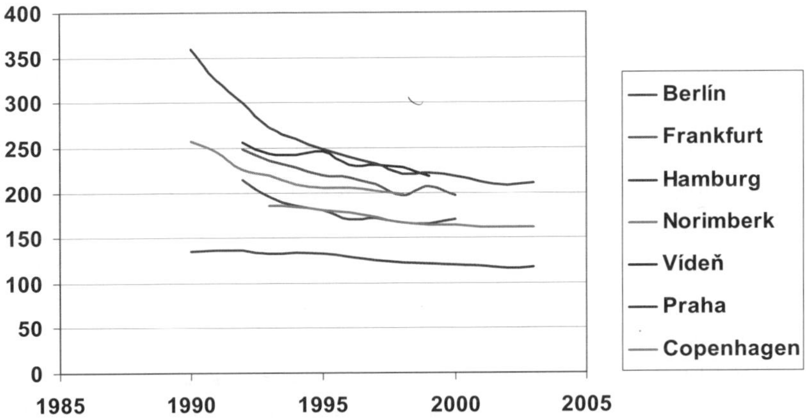

Figure 2. The average water consumption in some towns in Europe ($l*cap^{-1}*day^{-1}$)

The water supply system is always designed for water consumption during a given period of time, providing that it is possible to estimate and predict crucial factors such as development of the population, purchasing power, technical and technology levels as well as political situation in the given region; as such, it is negatively affected by contingent (unpredictable) events. Water management companies therefore strive to learn about methods how to regulate water consumption. Water consumption is likely to be reduced significantly during major negative events, e.g. environmental incidents. Reducing water consumption, however positive it may be in terms of maintaining the quantity of quality drinking water, is not a positive trend for water management companies, and it often leads to pressures on increasing the water price. Studies have been published on water demand regulation in line with the adjusted price and price plans. Variable financing makes it possible to set the water price more sensitively to make it acceptable, available, while ensuring that higher prices can be charged for high water consumption. It is evidently quite difficult to establish an ideal balance in the price of this indispensable resource for people. In a number of countries, drinking water prices are at least partially regulated by the government, and water prices are often subsidized by municipal and regional administration bodies. In the Czech Republic, water prices are still regulated by the government. However, better financial performance has been recorded by organizations in which the city or region exercises a certain influence, or which are subsidized by cities and regions.

3. European trend in the development of water consumption

- Reduced water consumption in the 1980s and 1990sStagnation in water consumption after 2000 Price of water Control consumption (e.g. Zurich or Madrid models).*Zurich model*1. payment for water supply commitment (permission to connect a house to the water supply system)2. payment for each water metering device, depending on the device or connection capacity3. Payments according to the contracted quantity per m^3
- *Madrid Model* Three basic water price plan types:1. by quantity consumed – induces people to save and reduce water consumption 2. a lump sum fee + measurements of the quantity consumed– pay ments are made also by people who use their own water sources but are also connected to the water supply system
- 3. the lump sum fee includes a portion of water consumption, allowing customer manipulation,establishing a social sentiment. As a result, it actually increases water consumption, the customers stop monitoring their consumption.

4. Situation in the Czech Republic

- State-regulated price
- Price for water plus sewerage + VAT
- Single-component price ($P * m^3$)
- Two-component price (flat fee + $P * m^3$), the flat fee depends on water meter size
- Attempts to regulate water consumption

Price of water

Water + sewerage charge + 5% VAT

Year 2004: 21,95 + 19,48 = 41,43 CZK/m^3

In EURO: 0,7+0,6 =1,3 €/m^3

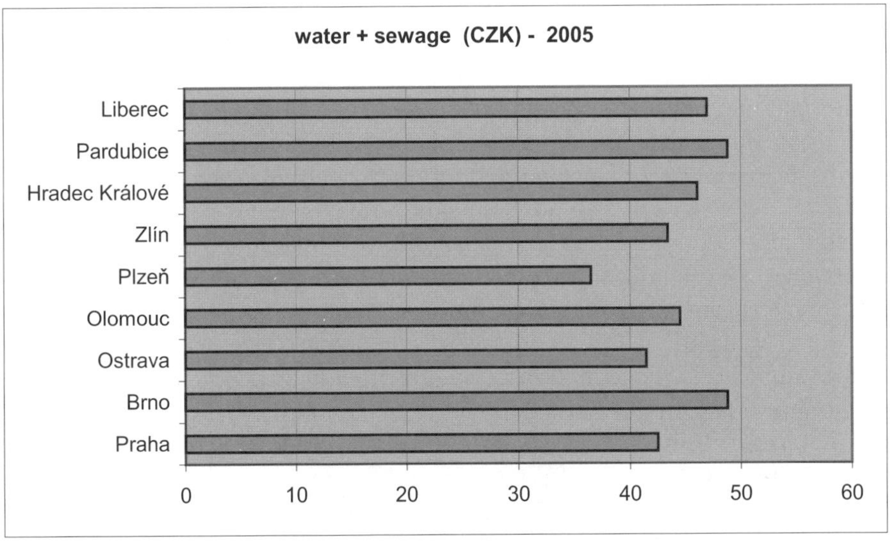

Figure 3. Price of water in Czech towns

5. Calculating of water consumption

- Specific consumption trends
- Population trends
- Trends in the industry, agriculture and service sector
- Housing furnishing standard
- Magnitude of losses

6. Efficiency of water works systems

6.1. WATER DISTRIBUTION

System of feed and supply water mains.

Resource – point of consumption optimisation.

6.2. WATER DISTRIBUTION PLAN

(pressure zone, local water mains, and group water mains, regional water works system)W for R = PW + $W_{received}$ – $W_{supplied}$

(Water for realization)

- W for R = WB + WNB
  (Water billed + water no billed)
- WB = WBH + WBIAO
  (Water billed in households), (Water billed.in industry, in agriculture, others)
  WNB = OD + damage + leakage
  (own demand)
- Capacity of water mains and water works facilities.
- Water quality as a function of transportation speed (detention time).

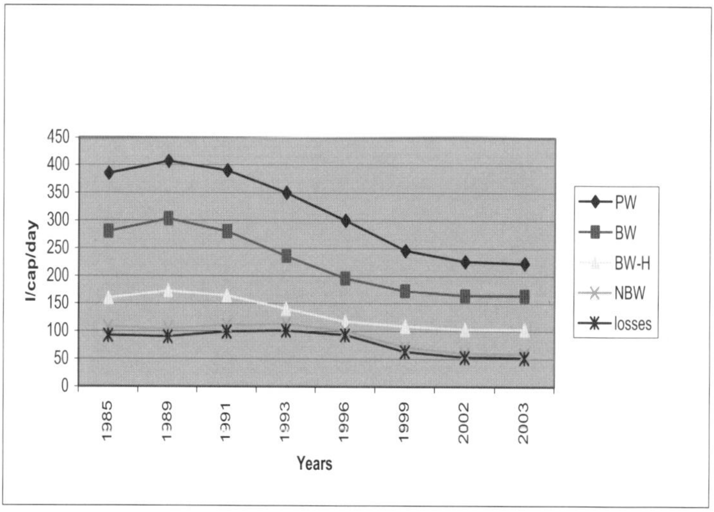

Figure 4. Water demand and efficiency of water work systems in the Czech Republic in 1985-2003

7. Reducing losses and water quality assurance

The main parameters which should be considered:

- Separate pressure zones (pressure – leaks reduction optimisation),
- measuring Q,
- regional invoicing,
- monitoring water leakspipe repairs large scale reconstruction
- permanent metering in very problematic districts

Refurbishment of water mains, reconstruction, no-dig technologies

- Recuperation spraying (PUR, cement lining)
- Lining of water mains pipes (on-site hardened materials, self-inflating helical foils,)
- New pipe retraction

Technical development

- Pipe diagnostics
- Reduction and control valves
- Technologies of refurbishing pipes and objects
- Automation, command transmissions
- Simulations, calculations, water mains database (GIS)
- Quality monitoring

Financing

- Infrastructure owners' money – local and city councils (repairs, investments)
- Operators' money (repairs)
- State influence – subsidies of Agriculture, Environment and Local Development ministries
- EU programmes, cohesion funds, co-funding conditions
- PPP programmes (Public Private Partnership)

8. Conclusion

The system of water supply is a very sensitive issue. When considering a potential attack on this system, it would lead to extensive losses and consequences.

Measures that are used to reduce water loss will also be adopted in the case of potentially jeopardized water supply systems.

Priorities

- Close the system,
- Separate affected areas,
- Prefer water supplied in bottles to temporary water supply system operation,
- Develop local water supply systems,
- Ensure that water pipes do not leak.

In order to reduce the potential risks and damage, it will be necessary to:

- Prepare procedures for supplying water in critical situations,
- Know alternative sources of water (in terms of location – place, altitude, collection areas, capacity),
- Prefer local water supply systems,
- Make it possible to divide consistently each of the existing large systems

Acknowledgement

This research was supported by Czech Ministry of Education, Youth and Sport research program MSM 6840770002.

References

Cihakova, I.: Water Losses in Water Distribution Systems, Workshop Brno 2003.
Cihakova, I.: Criteria for the Evaluation Water Losses. VI International Conference in Zlin, SOVAK, 2002.
Farley, M., Trow S.: Losses in Water Distribution Networks, IWA Publishing 2003.
Patera, A., et al.: Floods: forecasting, streams, landscape. Prague 2002. ISBN 80-01-02561-6.

USE OF DISTRIBUTION SYSTEM WATER QUALITY MODELS IN SUPPORT OF WATER SECURITY

WALTER M. GRAYMAN*, PhD, P.E.
W.M. Grayman Consulting Engineer, Cincinnati, Ohio, USA 45215

Abstract. Hydraulic and water quality models of water distribution systems are routinely applied for operational and design purposes. In the past few years, these models have been increasingly used in issues related to water security. This paper provides a general overview and history of water quality modeling in distribution systems. It also discusses how these models are being used to address water security issues in the area of planning and vulnerability assessment, design of monitoring networks, historical reconstruction of waterborne outbreaks, and real-time assessment, emergency response and remediation during a contamination event. Research and development needs associated with water quality modeling in support of water security are also discussed.

Keywords: water quality modeling; water security; distribution system

1. Water quality modeling of distribution systems

1.1. HISTORICAL BACKGROUND

Water quality distribution system models are used to simulate the behavior of a constituent as it moves with the water in a distribution system. Such models utilize flow and velocity information generated by a hydraulic model of the distribution system. Though hydraulic models date back to the 1930s (Cross, 1936), water quality models are relatively

---

**To whom correspondence should be addressed: Walter M. Grayman, Consulting Engineer, 321 Ritchie Avenue, Cincinnati, Ohio, USA, 45215; email: grayman@fuse.net*

J. Pollert and B. Dedus (eds.), Security of Water Supply Systems: From Source to Tap, 39–50.

new developments that have evolved only in the past quarter of a century.

Wood (1980) developed the first water quality model in a study of slurry flow in a pipe network. In a generalization of this formulation, Males, et al. (1985) used simultaneous equations to calculate the spatial distribution of variables that could be associated with links and nodes such as concentration, travel times, costs and other variables. By the mid-1980s, water quality models were developed that incorporated the dynamic behavior of water networks (Grayman, et al., 1988). The usability of these models was greatly improved in the 1990s with the introduction of the public domain EPANET model (Rossman, 2000). This led to the development of other Windows based commercial water distribution system models that are in use today. In more recent years, research and development has centered on modeling chemical reactions, integrating models with other systems such as geographic information systems, and application of optimization techniques to assist in model calibration, and system operation and design.

1.2. WATER QUALITY MODELING PROCEDURES

A hydraulic model of a distribution system is constructed by representing some or all of the pipes in the system and facilities such as pumps, tanks, and valves (Walski et al., 2003). The system is represented as a series of links and nodes. Some of the information that is required by the model includes pipe lengths, diameters and roughness factors, nodal elevations and demands, and specific information describing pumps, tanks and valves. Following the construction of the model, it must be properly calibrated so that it adequately reproduces the observed behavior in the actual system. Since system demands (and consequently the flows in the network) vary over the course of a day, extended period simulation (EPS) models are needed to simulate distribution system behavior under time varying demand and operational conditions. Water quality models utilize the results of the hydraulic model as input. They can be used to simulate chemical concentration, calculate water age, or to trace the percentage of water coming from a designated source (Clark and Grayman, 1998).

2. Water security applications

2.1. PLANNING AND VULNERABILITY ASSESSMENT

What is the impact of a contaminant that is purposely or accidentally injected into a distribution system? This depends upon the quantity, duration, location and toxicity of the chemical that has been injected into the system. A hydraulic/water quality model such as EPANET can be used to simulate the movement and behavior of the contaminant in the distribution system and to determine its impacts over time. Figure 1 shows the simulated chemical concentration throughout the network 15 hours after a chemical has been injected into the transmission line from the river source over a period of eight hours.

In Figure 2, the temporal variation in chemical concentration at a node in the central part of the distribution system is shown. This information can be used to estimate the level of chemical exposure for a person located at that site. Information such as that shown in Figures 1 and 2 has been used in planning and vulnerability assessments to determine the impacts of a specific contaminant event or to determine the general vulnerability of parts of a distribution system to a contamination event.

PipelineNet and TEVA (Threat Ensemble Vulnerability Assessment) are two recently developed planning tools that utilize hydraulic/water quality models to assess the impacts of potential contamination events.

(Bahadur et al., 2003). For example, it can be used to simulate the movement of contaminants through a distribution system and to display the results in order to show its impact upon sensitive facilities such as hospitals and schools that are contained in the GIS. TEVA is a probabilistic framework for assessing the vulnerability of a water utility to a large range of contamination attacks (Murray et al., 2004). Monte Carlo simulations generate ensembles of scenarios, and statistics are analyzed to explore the feasibility of scenarios, identify vulnerable areas of the water distribution network, and analyze the sensitivity of the model to various parameters.

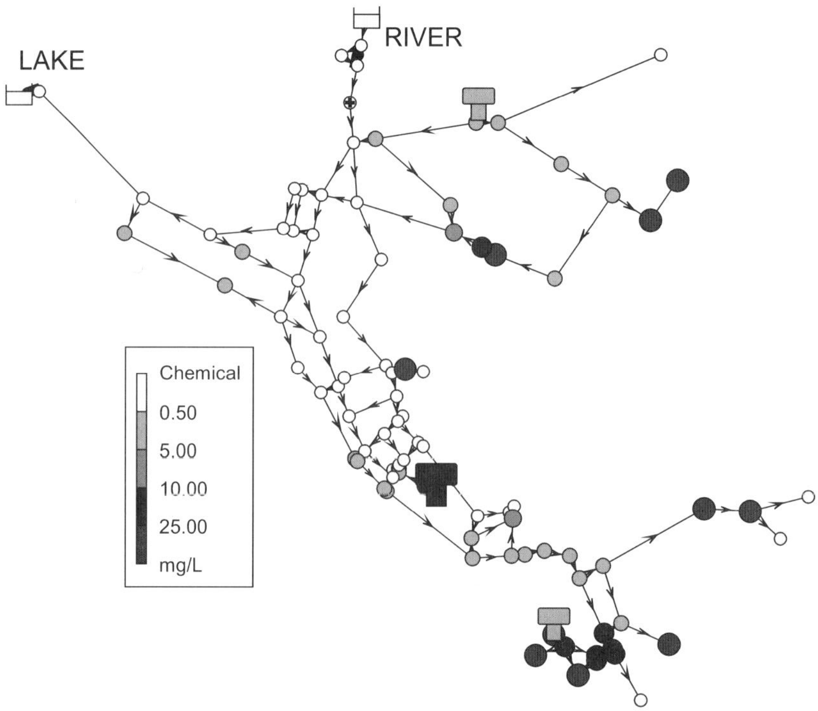

Figure1. Chemical concentration in network 15 hours after injection

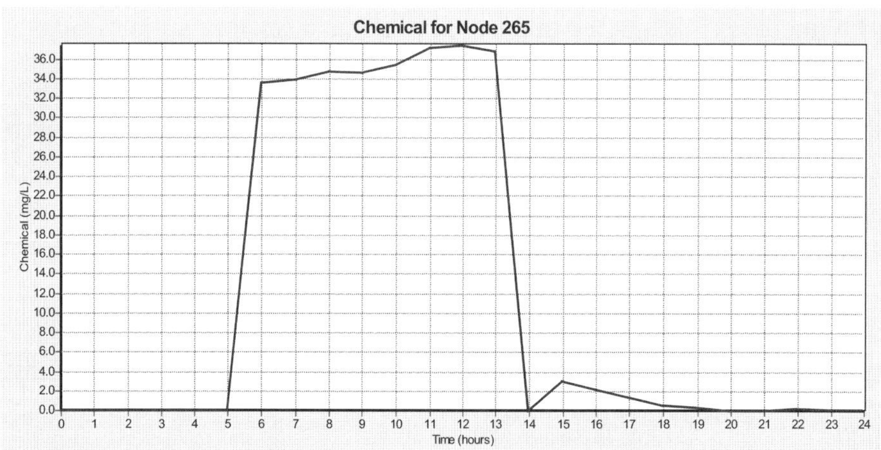

Figure 2. Temporal variation in chemical concentration at node

2.2. DESIGN OF MONITORING NETWORKS

On-line continuous monitors have been a major focus as a means of detecting a contaminant in the distribution system. In addition to selecting appropriate sensors, an important area of investigation has been the selection of optimal locations for monitors in order to provide the best capability for detecting contaminants. Various optimization and heuristic tools have been postulated to assist in the selection of monitoring locations in the distribution system. Each of the methods combine a hydraulic/water quality model with a search technique and some form of objective function to gage the effectiveness of a particular set of monitors and to select the locations that best fulfill the objectives.

Lee et al. (1991) proposed a method for locating monitors based on the concept of coverage which is defined as the percentage of total demand that is sampled by a set of monitors. Berry et al. (2004) used an integer programming optimization technique to place a limited number of perfect sensors in the pipes or junctions of a water network so as to minimize the expected amount of damage to the public before detection, assuming the attack occurs on a typical day. Watson et al. (2004) used a mixed-integer linear programming model for the sensor placement problem over a range of design objectives. Ostfeld and Salomons (2004) linked the EPANET model and a genetic algorithm in an overall framework for optimally locating monitoring stations, aimed at detecting deliberate external injections into water distribution system nodes. PipelineNet (Bahadur et al., 2003) links EPANET and GIS technology and has been used to guide a manual placement of monitors in order to fulfill certain general criteria.

Grayman et al. (2005) illustrated the use of the Ostfeld-Salomons model to select the optimal locations for sensors in a distribution system and to compare results to traditional ad hoc methods for sensor selection. Figure 3 illustrates the sensor locations selected by the optimization model based on one set of assumptions. In this case, it was assumed that the pollutant could be injected at any single node of the distribution system at any time, all with the same injection probability and that three sensors could be placed in the distribution system. Other assumptions were made that reflected the lethality, duration and quantity of chemical injected, the sensitivity of the sensor, and the response time following detection. The model suggests placing monitors at nodes 143, 181, and 213, resulting in a probability of about 44% that the contaminant will be detected prior to the consumption of more than 25 gallons at a concentration higher than 1 mg/L. As illustrated, the optimal sensor locations were spaced throughout the system (north, central and south) and generally reflected the locations chosen using more traditional inspection methods.

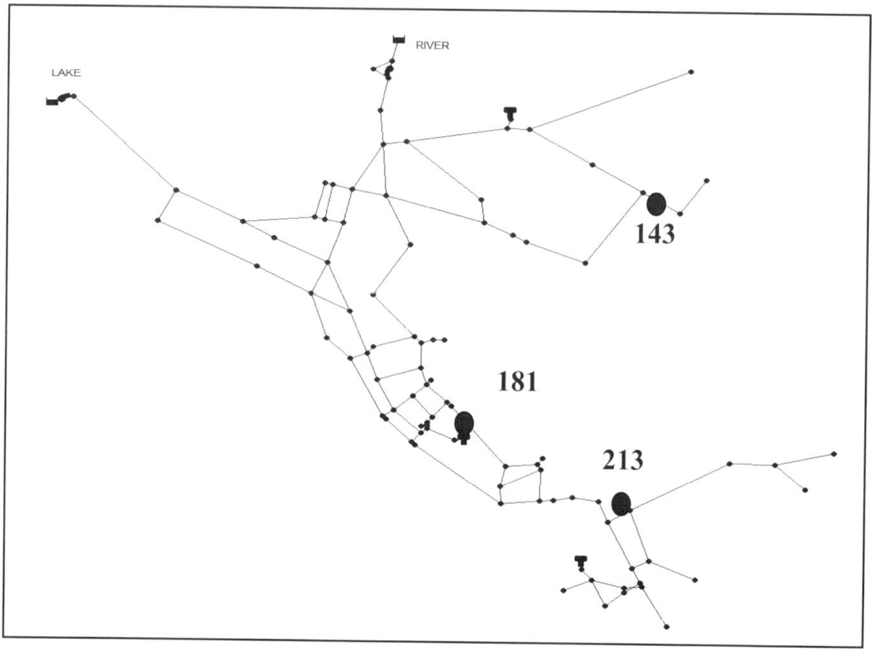

Figure 3. Monitoring locations selected by optimization routine

2.3. CONTAMINATION SOURCE DETERMINATION

Hydraulic/water quality models can be used in a forensic mode to determine the likely source of a contaminant that has been observed in a distribution system. The models could be applied in real-time to determine and characterize a source or could be used following an event to understand its origin. Trial and error application of a model represents the least sophisticated method for source determination. In this mode, the model is run repeatedly with differing sources until one is found that explains the observed data. However, this process can be very time consuming and tedious, and may not result in a unique solution.

Alternatively, mathematical optimization techniques can be applied that efficiently search and evaluate alternative source location/characteristics and select the one that best fits the observed data. In the distribution system, the source identification problem is a relatively new one that is still an area of research. This form of analysis is referred to as contaminant source determination, back tracing or an inverse solution. In a recent study, Laird et al. (2004) formulated a nonlinear program to estimate the time and location of contamination sources, using time varying concentration data from an installed sensor grid. In this program, the inverse problem is formulated as a least squares minimization subject to the differential and algebraic constraints of the network water quality model. Unknown time dependent injection terms are introduced at every network node, and the solution of the nonlinear program provides an estimate of the time and location of contamination sources.

2.4. REAL-TIME ASSESSMENT, EMERGENCY RESPONSE & REMEDIATION

When a contamination event occurs and it is detected, a rapid response is needed to minimize the impact of the event. A properly formulated and current hydraulic/water quality model of the distribution system can be an important part of that response. Such a model can be used to assess and predict the movement of the contaminants through the distribution system, determine the source of the contamination (if not known), provide information on how the movement of the contaminant can be controlled, and as part of the development of a clean-up strategy.

An important part of any real-time application of a hydraulic/water quality model is the ability to know the status of the water system at any point in time in order to serve as a starting point for modeling. An excellent source of information on system status is the SCADA (Supervisory Control and Data Acquisition) systems in larger water systems. SCADA continually monitors flow, pressure and water level data at key locations. There have been several demonstration projects, primarily by vendors of hydraulic modeling software and SCADA systems, to establish the feasibility of linkages between models and SCADA.

Joshi et al. (2004) describe a program called SCADAConnect to facilitate communication between Bristol Babcock's OpenEnterprise SCADA system and the WaterCAD modeling software package. This interconnection can be used for model calibration, real-time system control, and emergency response planning and forensic analysis. Horner and Ingeduld (2003) describe a program for connecting the MIKE NET hydraulic modeling package and SCADA that provides the capability to model the water distribution system in real-time thus integrating on-line modeling and monitoring of the system. The on-line module operates on top of the SCADA system and performs an on-line analysis of the system. The model results are stored back into the SCADA database for later viewing of the detailed model results.

Walski (2001) describes several scenarios of how a hydraulic/water quality model could be used in real-time during an emergency. He postulates how the model could be used to determine which valves should be closed and pumps that should be turned on or off in order to contain the contaminated water. He also demonstrates the use of the model to design a flushing program to get rid of the contaminated water with minimal impact on the consumer and the environment.

In general, real-time use of modeling is in its early formative stage and additional research, development and testing is required to make it an everyday tool.

2.5. HISTORICAL RECONSTRUCTION OF WATERBORNE OUTBREAKS

There have been many cases of contamination of water systems that have resulted in waterborne outbreaks associated with consumption of thewater. A thorough understanding of past contamination events and how and why they occurred and evolved is important for many reasons (Grayman et al., 2004):

- Remediation and correction of the specific causes of the contamination event
- Identifying the full extent of damage and injury
- Identifying liability for the event
- Providing general knowledge that will reduce the likelihood of contamination events and the vulnerability of water systems to contamination

The process of understanding the contamination event is referred to as historical reconstruction of the event. Distribution system hydraulic/water quality models have served as key tools in the quantitative reconstruction of a number of major waterborne events.

Events included both short-term acute contamination events and long-term chronic events.

Short-term acute contamination events generally result from the introduction of a pathogen into the water system that results in illness or death within a few days or weeks. An early example of such an event was the cholera epidemics in Britain in the 1850s that led to the benchmark methods used by John Snow to identify the specific cause and location of the contamination (Snow, 1855). Significant waterborne disease outbreaks in North America that have been the subject of historical reconstruction include an outbreak of E. coli serotype 0157:H7 in Cabool, Missouri in 1989 (Geldreich et al., 1992), a Salmonella outbreak in Gideon, Missouri in 1993 (Clark et al., 1996), widespread increase of diarrhea in Milwaukee, Wisconsin linked to the organism Cryptosporidium (Fox and Lytle, 1996), and an outbreak in Walkerton, Ontario in 2000 associated with waterborne E. coli O157:H7 and/or Campylobacter (Bruce Grey Owen Sound Health Unit, 2000).

Long-term chronic contamination events generally result from the introduction of an industrial chemical into a surface water or groundwater source over a period of years. The book, A Civil Action (Harr, 1995), details the investigation and legal activities associated with the contamination of the groundwater supply serving Woburn, Massachusetts, U.S. by industrial chemicals and its subsequent health impacts in that city. In this study, groundwater models were applied to estimate the movement of the pollutants to the wells and simplified, steady-state models of the water distribution system were applied to describe their likely movement through the water distribution system to identify the extent of customers that may have been exposed to the contaminated water (Murphy, 1986). In recent cases, more sophisticated modeling capabilities have been employed. Long-term (multi-year) simulations of the movement of trichloroethylene (TCE) in the distribution system were simulated in Phoenix and Scottsdale (Harding and Walski, 2000). The Agency for Toxic Substances and Disease Registry (ATSDR) modeled the movement of water from well sources through the Dover Township, NJ distribution system over the period from 1962 to 1996 in conjunction with epidemiological studies investigating clusters of childhood leukemia and nervous system cancers (Maslia et al., 2001). Uncertainty was incorporated in the historical reconstruction of the movement of TCE and the perchlorate anion through the distribution system in Redlands, California (Harding and Grayman, 2002).

3. Conclusions

Hydraulic and water quality modeling of distribution systems is quite widespread and in many applications such as fire flow analysis, design and master planning can be considered routine. However, applications in the area of water security are still on the cutting edge and certainly not yet routine. In the water security area, research and development are actively ongoing in two general areas: (1) assuring that the basic hydraulic/water quality models are sufficiently detailed and robust to support the needs of this form of analysis; and (2) development of the various methodologies and code needed to support the ancillary security related analyses described in this paper.

In the first general area, issues such as the needed level of detail, the required type and extent of calibration, and representation of the stochastic demands are being addressed. Calibration methods involving tracer studies and continuous monitors are being developed. Various methods of measuring and more accurately representing demands are and will be studied. This research is oriented towards insuring that the models that are applied in the water security area are sufficiently accurate to support the important decisions and actions that may be taken based on model results.

In the second general area, a wide range of theoretical and applied research has been accomplished, is underway, and will be performed in the future to address water security needs. The full range of such research will be limited only by the imagination of the researchers and the funding by governmental and private organizations.

References

Bahadur, R., Samuels, W.B., and Pickus, J., 2003, Case Study for a Distribution System Emergency Response Tool, Awwa Research Foundation, Denver, CO.

Berry, J., Hart, W. Phillips, C., and Uber, J., 2004, A general integer-programming-based framework for sensor placement in municipal water networks, Proc. World Water & Environmental Resources Congress, EWRI-ASCE, Reston, VA.

Bruce Grey Owen Sound Health Unit, 2000, The Investigative Report of the Walkerton Outbreak of Waterborne Gastroenteritis; http://www.publichealthgreybruce.on.ca/_private/Report/SPReport.htm

Clark, R.M. and Grayman, W.M., 1998, Modeling Water Quality in Drinking Water Systems, AWWA, Denver, CO.

Clark, R.M., Geldreich, E.E., Fox, K.R., Rice, E.W., Johnson, C.H., Goodrich, J.A., Barnick, J.A., and Abdesaken, F., 1996, Tracking a Salmonella Serovar Typhimurium Outbreak In Gideon, Missouri: Role of Contaminant Propagation Modeling, J. Water Supply Research and Technology - Aqua, 45(4): 171-183.

Cross, H., 1936, Analysis of Flow in Networks of Conduits or Conductors, Univ. of Ill. Eng. Experiment Station Bulletin 286, Urbana, IL.

Fox, K.M. and Lytle, D.A., 1996, Milwaukee's crypto outbreak: investigation and recommendations, J. American Water Works Association, 88(9): 87-94.

Geldreich, E. E., Fox, K. R., Goodrich, J. A., Rice, E.W., Clark, R. M., and Swerdlow, D. L., 1992, searching for a water supply connection in the Cabool, Missouri disease outbreak of Eschericia coli O157:H7, Water Research, 26:1127-1137.

Grayman,W.M., Ostfeld, A., Salomons, E., 2005, Red team-blue team exercise for locating monitors in distribution systems. Proc. World Water & Environmental Resources Congress, EWRI-ASCE, Reston, VA.

Grayman, W.M., Clark, R.M., Harding, B.L., Maslia, M, Aramini, J., 2004, Reconstructing historical contamination events, Water Supply Systems Security, L. Mays, ed., McGraw-Hill, New York, pp 10.1-10.55.

Grayman, W.M., Clark, R.M., and Males, R.M., 1988, Modeling distribution-system water quality dynamic approach, J. Water Resources Planning and Management, ASCE, 114(3): 295-312.

Harding, B.L., and Walski, T.M., 2000, Long time-series simulation of water quality in distribution systems, J. Water Resources Planning and Management, ASCE, 126 (4): 199-209.

Harding, B.L., and W.M. Grayman. 2002. Historical reconstruction of contamination in a distribution system incorporating uncertainty, 12th Annual Conference of the International Society of Exposure Analysis, 14th Conference of the International Society for Environmental Epidemiology., Vancouver, B.C.

Harr, J., 1995, A Civil Action. Random House, New York.

Horsner, M., and Ingeduld, P, 2003, Real time analysis for early warning systems and contingency planning. 2nd North American DHI Software Conference.

Joshi, P., Walski, T., Gandhi, S., Andrews, J.A., and Newswanger, C.F., 2004, Case study: linking Bristol Babcock's SCADA systems to WaterCAD, a water distribution modeling tool. Proc. AWWA DSS Conf., Denver, CO.

Lee, B.H., Deininger, R.A., and Clark, R.M., 1991, Locating monitoring stations in water distribution systems, J.AWWA, 83(7):60-66.

Laird, C.D., Biegler, L.T., van Bloemen Waanders, B.G., and Bartlett, R.A., 2005, Time dependent contamination source determination: a network sub domain approach for very large water networks. J. Water Resources Planning and Management, ASCE, 131(2): 125–134.

Males, R.M., Clark, R.M., Wehrman, P.J. and Gates, W.E., 1985, Algorithm for mixing problems in water systems, Jour. of Hyd. Eng., ASCE, 111(2): 206-219.

Maslia, M. L., Sautner, J. B., Aral, M. M., Gillig, R. E., Reyes, J. J., and Williams, R. C., 2001, Historical Reconstruction of the Water-Distribution System Serving the Dover Township Area, New Jersey: January 1962–December 1996, Agency for Toxic Substances and Disease Registry, Atlanta, GA.

Murphy, P.J., 1986, Water Distribution in Woburn, Massachusetts. Pub. 86-1. The Environmental Institute, University of Massachusetts, Amherst, MA.

Murray, R., Janke, R., and Uber, J., 2004, The threat ensemble vulnerability assessment (TEVA) program for drinking water distribution system security. Proc. World Water & Environmental Resources Congress, EWRI-ASCE, Restion, VA.

Ostfeld A., and Salomons E., 2004, A stochastic early warning detection system model for drinking water distribution systems security", Proc. World Water & Environmental Resources Congress, ASCE, Reston, VA.

Rossman, L.A., 2000, EPANET Version 2 Users Manual, Drinking Water Research Division, USEPA, Cincinnati, OH.

Snow, J., 1855, On the Mode of Communication of Cholera. John Churchill, London (UK), 191 pp.

Walski, T.M., Chase, D.V., Savic, D.A., Grayman, W.M., Beckwith, S., and Koelle, E., 2003, Advanced Water Distribution Modeling and Management, Haestad Press, Waterbury CT.

Walski, T. M. 2001. Water distribution models for emergency response. Proc. Water Security Summit, Haestad Press, Waterbury, CT.

Watson, J.P., Greenberg, H.J., and Hart, W.E., 2004, A multiple-objective analysis of sensor placement optimization in water networks. Proc. World Water & Environmental Resources Congress, EWRI-ASCE, Reston, VA.

Wood, D.J., 1980, Slurry flow in pipe networks. J. Hydraulics, 106(1): 57-70. ASCE, Reston, VA.

VULNERABILITY OF WATER DISTRIBUTION SYSTEMS TO LEAKAGE

VLADIMIR HAVLIK*
Czech Technical University in Prague, Faculty of Civil Engineering, Thakurova 7, 166 29 Prague 6, Czech Republic

Abstract. Vulnerability of Water Distribution Systems to Leakage is defined using standard terminology. The use of Water Balance Method illustrates a basic approach to leakage. Moreover, the suitability of various performance indicators is discussed and illustrated using both technical as well as financial data. Methodology to evaluate leakage (e.g. DMA, SCADA) provides the operators with a suitable means how to decrease the leakage. Field measurements of hydraulic parameters and mathematical modelling of Water Distribution Systems may substantially help to quantify leakage and to evaluate the best practice for integrated leakage reduction.

Keywords: definitions / water balance method / performance indicators / real losses / leakage assessment / mathematical modelling

1. Introduction

The first Water Security Summit, which was held in Hartford, Connecticut, USA on December 3-4, 2001, tackled all aspects of vulnerability of Water Distribution Systems (WDS). Every WSD is vulnerable to possible physical, chemical, and biological activities that could interrupt water supply and result in contamination. The paper deals with leakage that is inherent physical property of any WSD. In the beginning the leakage problem is defined, and then it is illustrated how it is possible to evaluate leakage and finally how mathematical models can be used to assess the water system vulnerability.

**To whom correspondence should be addressed: Vladimir Havlík Czech Technical University in Prague,Faculty of Civil Engineering,, Thakurova 7, 166 29 Prague 6, Czech republic, vhavl@hydroprojekt.cz*

J. Pollert and B. Dedus (eds.), Security of Water Supply Systems: From Source to Tap, 51–63.

2. Definition of leakage

Every WDS is vulnerable to leakage. The leakage is universal problem and in order to cope with it, the following approach should be taken into account:

1. Identify the causes of leakage.
2. Use appropriate methodology and tools to quantify the leakage and to reduce it.

It should be pointed out that the leakage might be classified from several points of view. However, from the practical viewpoint, it is useful to distinguish between pipe bursts (PB) and background leakage (BL). PB occurs in the water mains at some particular pipe cross-sections and/or joints and sometimes is called reported (visible) leaks. The amount of lost water is great and PB manifests itself after relatively short time on the street surface. Water flows over the surface and frequently a destruction of surface as well as huge soil erosion occur. It often takes tenths of minutes than the water supply is stopped and the repair can start. Farley and Trow (2003) point out that the lost volume of water depends on awareness, location and repair times. While awareness time is the time until the water company gets know about the leak, location time is needed for the leak location. Repair time needs the water company to repair the leak.

On the other hand, BL is hidden and it cannot be detected just by eye. BL is represented by numerous small leaks that often cause a significant proportion of total leakage. Considering flow rates caused by leakage, there are unreported bursts (UB) between these two groups of leakage.

The vulnerability of WDS to leakage depends on many factors and among others the following are important:

- Pipe material and age of pipelines.
- Protection of pipelines from corrosion, pitting effect, winter conditions. The method of jointing the pipes, ground conditions and soil type, as well as surface loadings is also important.
- Operation of WDS from hydraulic point of view (steady flow, quasi-steady flow, hydraulic transients). The higher is the pressure, the greater leakage is.

3. Methods for evaluation of leakage

Figure 1 shows general approaches how to evaluate leakage. Each approach may be applied separately; however, the best results are obtained with suitable combination of more possibilities

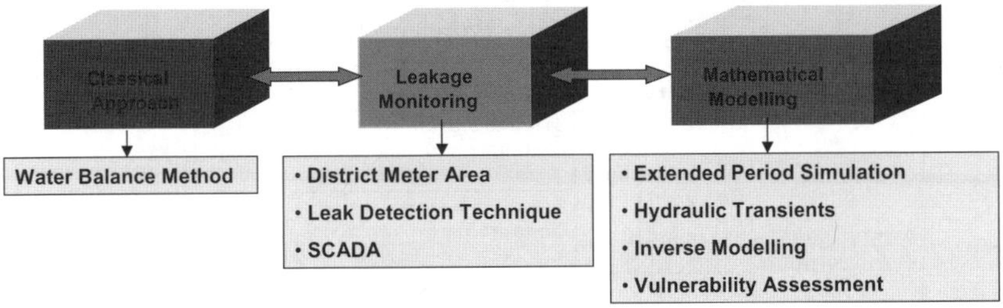

Figure 1. Approaches to Leakage Evaluation

3.1. WATER BALANCE METHOD

It is well recognised that the most important is reliable metering of all water volumes, i.e. the System Input Volume (SIV), which is entering the system, measurements of night flows and in particular metered consumption at customer's service pipes. Water Balance Method (WBM) is recommended to use within a time interval 1 year, which means that each component is expressed in volume of water per year. Lambert and Hirner (2000) prepared in IWA Report the international definitions and components of water balance components, see TABLE 1.

SIV - System Input Volume is the input to transmission mains or to WDS.
BMC - Billed Metered Consumption is the volume taken by registered customers for domestic, commercial and industrial purposes.
BUMC - Billed Unmetered Consumption is the volume derived e.g. from sample metering.
BAC - Billed Authorised Consumption is the sum of metered and unmetered consumption.
UFW - Unaccounted-for-Water (Non-Revenue Water) is the difference between SIV and BAC.
UNMC - Unbilled Metered Consumption.
UUNMC - Unbilled Unmetered Consumption.
UNAC - Unbilled Authorised Consumption is the sum of Unbilled Metered and Unbilled Unmetered consumptions.
OC - Own WDS and water treatment technology consumption.
UNAUTC - Unauthorised Consumption.
MI - Metering Inaccuracies.
AL - Apparent Losses - theft of illegal use and inaccuracies associated with production and customer metering.
LTDM - Leakage on Transmission and/or Distribution Mains.
LOST - Leakage and Overflows at Utility's Storage Tanks.
LSCCM - Leakage on Service Connections up to point of Customer metering.
RL - Real Losses are physical losses from pressurized system up to the point of customer metering.
WL - Water Losses is the sum of Apparent and Real Losses.
WLPC - Water Losses expressed in percentage to the System Input Volume.

Table 1. Water Balance Method

<table>
<tr><th colspan="5">WATER BALANCE COMPONENTS FOR TRANSMISSION SYSTEM AND WDS</th></tr>
<tr><td rowspan="9">System Input Volume

SIV
(m^3/year)</td><td rowspan="4">Authorised Consumption (m^3/year)
AC
AC=BAC + UNAC</td><td rowspan="2">Billed Authorised Consumption (m^3/year)
BAC
BAC=BMC + BUMC</td><td>Billed Metered Consumption
BMC</td><td rowspan="2">Revenue Water (m^3/year)
RW</td></tr>
<tr><td>Billed Unmetered Consumption
BUMC</td></tr>
<tr><td rowspan="2">Unbilled Authorised Consumption (m^3/year)
UNAC
UNAC=UNMC + UUNMC</td><td>Unbilled Metered Consumption
UNMC</td><td rowspan="7">Non-Revenue Water

i.e.

Unaccounted-for-Water

(m^3/year)

UFW
UWF=SIV - RW</td></tr>
<tr><td>Unbilled Unmetered Consumption
UUNMC</td></tr>
<tr><td rowspan="5">Water Losses (m^3/year)
WL
WL= AL + RL</td><td rowspan="2">Apparent Losses (m^3/year)
AL
AL= UNAUTO + MI</td><td>Unauthorised Consumption
UNAUTC</td></tr>
<tr><td>Meetering Inaccuracies
MI</td></tr>
<tr><td rowspan="3">Real Losses (m^3/year)
RL
RL= LTDM + LOST + LSCCM
RL = WL - AL</td><td>Leakage on Transmission and/or Distribution Mains LTDM</td></tr>
<tr><td>Leakage and Overflows at Utility´s Storage Tanks LOST</td></tr>
<tr><td>Leakage on Service Connections up to point of Customer metering LSCCM</td></tr>
</table>

The best practice for water losses management is to combine continuous water balance calculations (usually taken over a 12-month period) and night flow measurements on a continuous or "as required" basis. The WB calculation quantifies volumes of total water delivered into the WDS, authorized consumption (billed or unbilled), metered or unmetered) and water losses (real or apparent), see TABLE 1. Each component of the annual WB should be expressed as volume per year. The steps for calculation of Non-Revenue Water and Water Losses could be as follows:

1. Define System Input Volume SIV
2. Define BMC and BUMC and calculate BAC
3. Calculate UWF as SIV minus RW
4. Define UAC
5. Determine AC as the sum of BAC and UNAC
6. Calculate Water Losses as the difference between SIV and AC
7. Determine the Apparent Losses AL
8. Calculate Real Losses as Water Losses minus Apparent Losses
9. Assess components of Real Losses by best means available (night flow analysis, mathematical modelling etc) and check this volume with volume of Real Losses that was derived from previous step.

3.2. PERFORMANCE INDICATORS

The quality of service that is provided by Water Utilities (WU) to the customers is a key issue. Using performance indicators (PI) enables to assess the actual level of performance. WU must look for a high degree of efficiency and effectiveness (1). *Efficiency* measures the extent to which the resources of a WU are utilized optimally to produce the service. *Effectiveness* measures the extent to which the targeted objectives (specifically and realistically defined) are achieved.

The PI is a quantitative or qualitative indicator that is related to certain aspects to measure performances and standards of service so that the efficiency and effectiveness of the utility can be determined. Detailed review of the performance indicators can be found in Alegre et al. (2002a,b). Water Utilities should comply with a logical managing direction to satisfy the main philosophy: greater satisfaction of a greater number of consumers and concerned entities, with the best use of available resources.

Over the past 10 years, there was a great effort in defining, selecting and finding the best structure for PI. Performance Indicators should become an integral part of WU management strategy and programme for rehabilitation procedure. In addition to traditional Operational indicators, Quality of service indicators, Financial indicators, Water resources indicators and Physical indicators, some Additional indicators may by useful – network hydraulic reliability, mains residual service life, service connections residual service life, balance of costs and benefits, internal rate of return etc.

One of the most important issues is Real Losses. Traditional PI representing the Annual Volume of Real Losses can be expressed as:

- % of SIV
- Figure per length of mains per day or hour
- Figure per service connection per day or hour
- Figure per property per day or hour
- Figure per length of system per day or hour

However, Real Losses expressed as percentage of SIV are unsuitable for assessing the efficiency of management of WDS. This is because this PI is not able to consider of any of the key influences on Real Losses and differences in consumption influence the value of Real Losses expressed in % terms. The final recommendation is that the basic Technical Indicator for Real Losses (TIRL) should be the Annual Volume of Real Losses divided by the number of service connections (Nc), allowing for the % of the year for which the system is pressurized (w.s.p.). TIRL can be expressed as follows

TIRL = Current Annual Volume of Real Losses / Nc (1)

[Litres/service connection/day w.s.p.]

A more detailed description of TIRL values can be obtained by comparing TIRL value with a "best estimate" of Unavoidable Average Real Losses (UARL). UARL allows for local conditions of connection density, location of customer meters and average operation pressure.

The calculation of UARL (Litres/service connection/day) is based on the following form of equation

UARL = (A.Lm + B.Nc + C.Lp) . P/Nc (2)

(Litres/service connection/day w.s.p.)

The advantage is that there is a separate influence of Real Losses from the length of mains (Lm in km), number of service connections (Nc), total length of service connections from the edge of the street to customer meters (Lp in km), and the average pressure (P in meters of water column) when the system is pressurized. The appropriate values of A, B and C were determined using statistical analysis of international sets of data, Lambert and Hirner (2000). Hence the final recommended equation be as follows

UARL = (18*Lm + 0.8*Nc + 25*Lp) * P/Nc (3)

The difference between TIRL and UARL represents the maximum potential for future savings in Real Losses, when the system is pressurized. Moreover, a useful non-dimensional Infrastructure Leakage Index (ILI) represents a current operation regime and the overall condition and management of infrastructure.

ILI = TIRL / UARL (4)

Again, based on the statistical analysis, ILI can range from close to 1 above 10. Well-managed systems in very good conditions should exhibit ILI values close to 1,0.

4. Leakage monitoring

Leakage monitoring is one of the most important activities in order to detect, locate and quantify the leakage. The most appropriate approach is called District Meter Area (DMA) when the supply network is divided into suitable sub-networks (e.g. discrete pressure zones and/or other parts of WDS that can be separated by boundary valves etc.) of up to 3000 properties. The flow meter is installed to record flows into DMA. In order to determine Real Losses, the most representative are the night flows. Leak detection techniques such as acoustic loggers or correlate survey are recommended for regular leak localization. Permanent monitoring system such as SCADA can substantially help in permanent collecting information, data processing and a step towards the Real Time Control.

5. Mathematical modelling

Mathematical modelling of WDS has become a standard approach that can offer a lot of possibilities. Mathematical models are used not only for determination of flow rates and pressures (both for steady and Extended Period Simulation, fire flow analysis), water quality simulation (water age, source tracing, chlorine concentrations etc), but also for reliability analysis, Mays W. (1994), vulnerability analysis, Walski, M.T. (2001) and leakage analysis, EC SPRINT Project, Liggett, J.A. and Chen, Li-Chung (1994).

Practical application of computer models require a good knowledge of basic principles as well as suitable use of information technology, Havlik, V. and Ingeduld, P. (1997a,b). Modelling of leakage also requires the knowledge of relationship between the pressure and leakage rate, Germanopoulos, G. and Jowitt, P.W. (1989). The pressure reduction will reduce the leakage. The empirical relationship can be expressed as follows

$$Q_L = k\Delta P^N \tag{5}$$

in which k is constant and N is the power factor, whose mean values can vary between 1.13 and 1.15.

6. Case study

6.1. KOCANI PILOT

The leakage study under the Ecolinks program, Havlik et al. (2003), was conducted in the district city of Kocani, Macedonia. The layout of WDS model is shown in Figure 2.

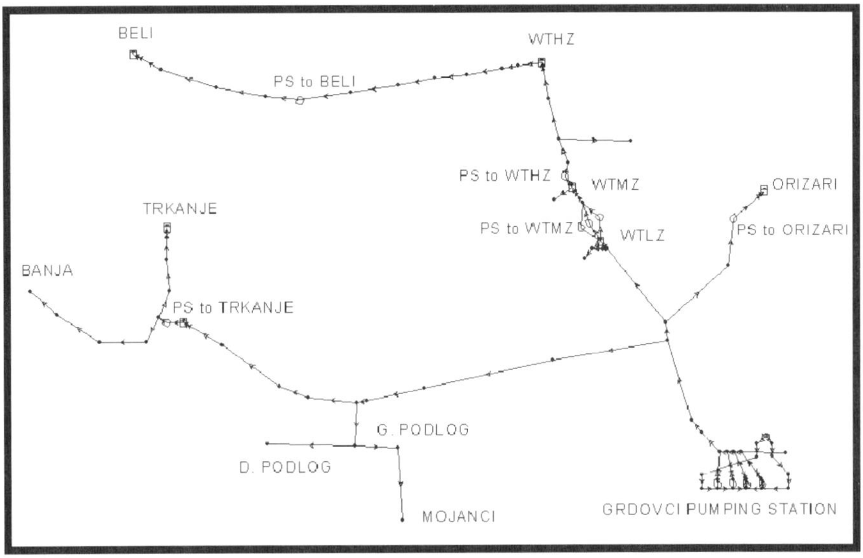

Figure 2. Layout of WDS in Kocani

The existing WDS in the city of Kocani provides the service not only for the city of Kocani, but as well as for neighbouring villages. There are several pumping stations and three main pressure zones. The principal pumping stations is situated in Grdovci.

Table 2 shows Water Balance Calculation for WDS in Kocani that has been applied since 1995. It can be seen that the trend for SIV is to increase; however, also Unaccounted-for-Water is still increasing. The value of 66% in year 2002 is alarming. The specific consumption of electric energy that is spent for pumping water is shown in Table 2 as well. Again, it can be seen the trend to increase these values.

An example of using simple as well as recommended PI from Chapter 3.2 is shown for the year 2002 in Table 3. Suitable expressing of real losses with the use is TIRL from Eqn. (1), UARL from Eqn. (3) and the Infrastructure Leakage Index ILI from Equation (4) show that its value for the WDS in Kocani is

Table 2. Water Balance Method used for the WDS in Kocani

1	2	3	4	5	6	7	8	9
Year	Water Produced	Billed Authorised Consumption	Water Losses	Water Losses WL/SIV	Water Losses	Consumption of Electric Energy	Specific Consumption of Electric Energy	Specific Consumption of Electric Energy
	WP	BAC	WL	WLPC	WL	EEC	SCEE/WP	SCEE/BAC
	(m^3)	(m^3)	(m^3)	(%)	(l/s)	(kWh/year)	(kWh/m^3)	(kWh/m^3)
1995	4 027 085	**2 218 557**	**1 808 528**	**44.9**	57.3	3 035 167	0.75	1.37
1996	4 711 292	**1 984 012**	**2 727 280**	**57.9**	86.5	3 406 762	0.72	1.72
1997	4 824 125	**1 986 122**	**2 838 003**	**58.8**	90.0	3 399 780	0.70	1.71
1998	4 883 194	**1 992 085**	**2 891 109**	**59.2**	91.7	3 486 917	0.71	1.75
1999	5 323 142	**1 961 175**	**3 361 967**	**63.2**	106.6	3 899 248	0.73	1.99
2000	5 572 949	**2 113 600**	**3 459 349**	**62.1**	109.7	4 049 173	0.73	1.92
2001	5 334 151	**2 075 140**	**3 259 011**	**61.1**	103.3	4 364 140	0.82	2.10
2002	5 558 676	**1 855 940**	**3 702 736**	**66.6**	117.4	4 529 388	0.81	2.44

ILI = 9.5. While well-managed systems in good conditions should receive ILI values close to 2.0, this rather high value indicates a great need for future rehabilitation programme.

Table 3. Water Loss Performance using Recommended PI

Length of Water Mains L_{hp} (km)	120	2 160	(L/day/P)
Number of Service Connections N_p	10 230	8 184	(L/day/P)
Length of Service Connections L_p	123	3 069	(L/day/P)
UARL	Eqn. (3)	13 413	(L/day/P)
UARL at average pressure P	P = 50 (m H_2O)	670 650	(L/day/P)
UARL in the same units as TIRL		65,6	(L/N_p/day)
TIRL	Eqn. (1)	620.7	(L/ N_p /day)
Ifrastructure Leakage Index ILI = TIRL / UARL	Eqn.(4)	9.5	(-)

The data sets from SCADA were used to determine the inflows into three pressure zones and into selected District Metering Areas. Example for the Low Pressure Zone (DMALZ) with a typical one-week variation of demands and night flow is shown in Figure 3.

The Real Losses in the city of Kocani represent about 80% of Real Losses in the whole WDS (RL in villages represent only 20% of RL). Moreover, the Real Losses represent about 67% of total losses (UFW) while Apparent Losses about 18% and UNAC the rest, i.e. 15%. Table 4 shows measured Real Losses

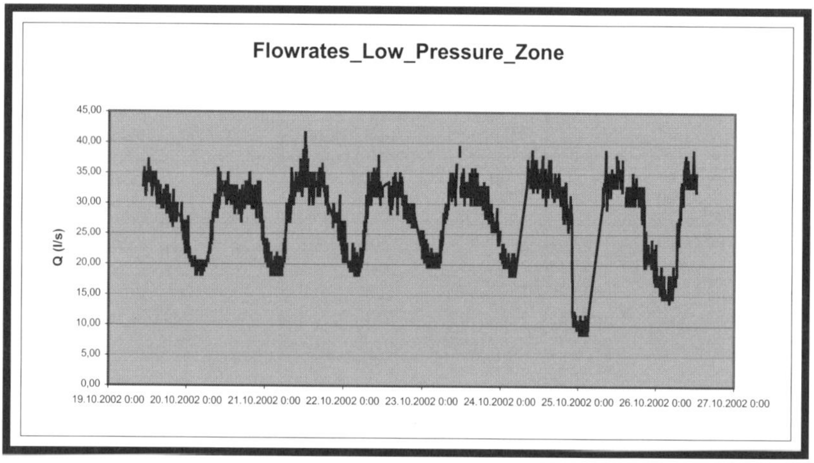

Figure 3. Typical Variation of Flow Rates into Low Pressure Zone

from night measurement in comparison with daily average demands. It can be pointed out that the Low Pressure Zone should be preferred not only from the point of view of measured Real Losses, but also when using PI that represents the Real Losses per service connections. The District Meter Area No 2 for the High Pressure Zone (DMAHZ2) exhibits the highest measured values of the Real Losses.

Execution of field measurements using SCADA and mathematical modelling using the computer model WaterCAD (Haestad Methods) enabled to perform leakage calculation. The leakage methodology was based on the Water Balance Calculation and suitable performance indicators.

Table 4. DMA Statistics from Measurements

DISTRICTS	Q_{RL} (L/s)	$Q_{av,d}$ (L/s)	$Q_{RL}/ Q_{av,d}$ (%)
DMALZ	20	32	62.5
DMAMZ	20	45	44.4
DMAHZ1	2.5	8	31.3
DMAHZ2	5	8	62.5
DMAHZ3	4.7	10	47.0
DMAHZ4	1	6	16.7
TOTAL HZ	13.2	32	41.3

6.2. REDUCTION LEAKAGE BY PRESSURE

Illustrative example using the network layout from the Low Pressure Zone in Kocani may illustrate how the reduction of pressure decreases the leakage rate. Leakage was calculated using model MikeNet supposing to be concentrated into nodes No 29, 58 and 68, see Figure 4. Once the pressure was decreased to 10 mwc in front of above-mentioned nodes, the leakage rate decreased by about 58%. For example, the original leakage rate in Node 29 was $Q_L = 1{,}16$ l/s and due to reduction in pressure the leakage rate decreased to $Q_L = 0{,}31$ l/s.

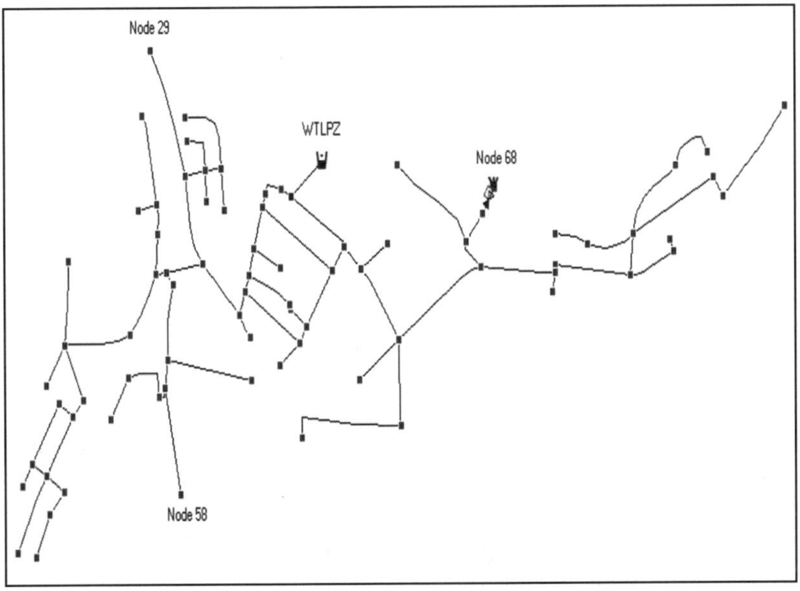

Figure 4. Layout for the Leakage

6.3. CONTAMINANT SIMULATION

The second illustrative example again with the use of the network layout from the Low Pressure Zone in Kocani may illustrate how the peak contaminant concentration from WTLPZ likely fates. The peak concentration has been introduced into WDS (either intentionally or accidentally) with c = 100 mg/l (triangular shape concentration within 2 hours). The reference nodes in Figure 5 were used to detect the movement of peak concentration of the contaminant.

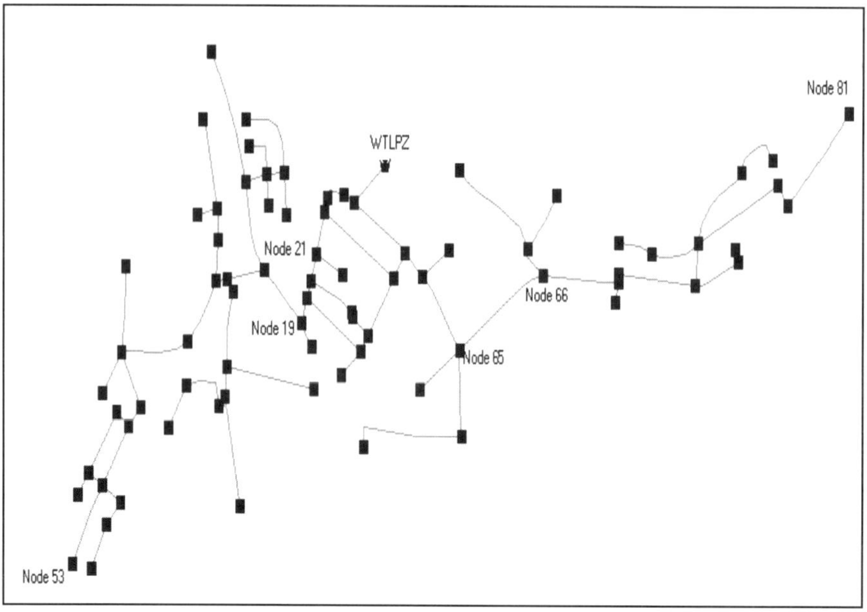

Figure 5. Layout for the Movement of Contaminant

The results with the use of model MikeNet can be summarised as follows: For example, the peak concentration (c = 74.5 mg/l) reached the Node 66 after 1 hour, while the Node 81 after 5 hours (c = 21.9 mg/l) and the Node 53 after 7 hours (c = 34 mg/l). The best remedial measure would be in this case quick closure of valves along the network. The second response could be flushing the system. As an example, flushing in Node 21 decreased the peak concentration from 36 mg/l down to 1,6 mg/l.

References

1. Alegre, H., Hirner, W., Baptista, J.M., Parena, R. (2002): Performance indicators for water supply services, Manual of Best Practice Series, IWA Publishing London, ISBN 1 900222 18 3 (150pp).
2. Alegre, H., Baptista, J.M., Coelho, S.T., Praca, P. (2002): Performance indicators for network rehabilitation, Proceedings from International Conference "Computer Aided Rehabilitation of Water Networks CARE-W", Hertz, K (editor), November 1st, 2002, Dresden, Germany, ISBN 3 86005-335-3.
3. EC SPRINT Project SP 257 (1997) Improving Network Management and Leakage Control in Five European Countries, BHRA Group 1997, Water Pipeline Systems.
4. Farley, M. and Trow, S. (2003) Losses in Water Distribution Networks. IWA Publishing, London, ISBN 1 900222 11 6.

5. Germanopoulos, G. and Jowitt, P.W. (1989) Leakage reduction by excess pressure minimization in a water supply network, Proc. Instn. Engrs, Part 2, pp.195-214.
6. Havlík, V. and Ingeduld, P. (1997) Hydraulic analysis of water supply and distribution networks. Proceeding 3rd International Conference on Water Pipeline Systems, The Hague, 13-15 May, pp. 395-405.
7. Havlík, V. and Ingeduld, P. (1997). Modelling tools for water supply networks, Proceedings of the European Water Resources Association, Relfsgaard&Karalis (eds), Balkema, Rotterdam, ISBN 90 5410 8975, pp. 445-452.
8. Havlik, V., Holy, P., Kushevski, T, Todorov, B.: Leak Abatement in Kocani Water Supply System, EcoLinks Challenge Grant Agreement CG6-MK-21, 2003.
9. Lambert, K., Hirner, W.: Losses from Water Supply Systems: Standard Terminology and Recommended Performance Indicators, IWA, October 2000.
10. Liggett,J.A. and Chen, Li-Chung (1994) Inverse transient analysis in pipe networks, Journal of Hydraulic Engineering, Vol. 120, No. 8, paper No 6420.
11. Mays, L. W. (1994) Methodologies for Reliability Analysis of Water Distribution Systems, M.H. Chaundry and L.W. Mays (eds.), Computer Modelling of Free-Surface and Pressurised Flows, 485-517, Kluwer Academic Publishers.
12. Walski,T.M. (2001) Application of water distribution system models, Proc. From Water Security Summit, December 3-4, 2001, Haestad Methods.

REAL TIME ANALYSIS FOR EARLY WARNING SYSTEMS

PETR INGEDULD*

DHI HYDROINFORM, Na Vrsich 5, Prague, Czech republic

Abstract. Water treatment and distribution systems are highly vulnerable to degradation of quality and reliability of supply as a result of many factors, natural, accidental, and intentional. Among the potential intentional factors that the utility manager and operators have to plan for is the introduction of toxic contaminants into the water supply or disruption of water service through sabotage of key components of the infrastructure. Rapid recognition of the nature and location of such occurrences is vital to protect the integrity of the water supply and safeguard the consumers from potentially harmful contaminants, determine appropriate changes in supply and treatment strategy, and ensure compliance with environmental regulations. The utility manager and the operations staff must be given the proper tools as well as be trained to identify an event, locate the extent and potential danger to the public, and be prepared to react in a proper and timely fashion. Rapidly developing sophisticated software and real time instrumentation and monitoring systems provide the tools to design and develop early warning monitoring systems and to increase the preparedness of the water utility to react to such unexpected events. Proper integration of state of the industry hydraulic modeling systems, geographical information systems (GIS) for the water distribution network, and the installation of a SCADA system for both water treatment plant and active element control as well as the monitoring of critical points within the distribution system will be an invaluable resource for the operator to react to an event (real time response) as well as to plan for possible future events (contingency planning).

Keywords: Real time modeling, SCADA, Online simulation, operations, hydraulic and water quality modeling, early warning systems, vulnerability of water supply, contingency planning, system integration

---

**To whom correspondence should be addressed: Petr Ingeduld, DHI Hydroinform, Na vrsich 5, Prague, Czech republic. ingeduld@dhi.cz*

J. Pollert and B. Dedus (eds.), Security of Water Supply Systems: From Source to Tap, 65–84.

1. Early warning monitoring systems

The goal of an early warning monitoring system is to reliably identify low probability /high impact contamination events (chemical, microbial, radioactive) in source water or distribution systems in time to allow an effective local response that reduces or avoids entirely the adverse impacts that may result from the client (Brosnan, 1999).

Requirements for the ideal early warning system:

- Provides warning in sufficient time for action
- Cost is affordable
- Requires low skill and training
- Covers all potential threats
- Is able to identify the source
- Is sensitive to quality at regulatory levels
- Gives minimal false positive or negative responses
- Is robust
- Is reproducible and verifiable
- Functions year-round

A key component of early warning systems is the availability of a mathematical model for predicting the transport and fate of the spill/contaminant so that downstream utilities can be warned. However, a water quality model should be regarded as a guide only to what will happen, and increased monitoring during spill events is often critical to verifying the model and determining the fate of the spill and the safety of the supply.

The number of intentional threats and acts of sabotage against water supply systems is relatively small, and hoaxes are very likely. Intentional threats include:

- Destruction of a water supply system. Destruction of parts of the system can happen by physical destruction or computer hacking. Cyber attacks against various computerized components of water supply systems include attacks against SCADA systems.
- Contamination of the system with chemicals, microbes, toxins, or radioactive compounds. Examples of chemical agents are nerve agents, cyanide, arsenic, and nicotine. The most likely points of attack for intentional contamination include post treatment storage reservoirs, distribution tanks, and water mains (Haimes et al., 1998).

1.1. REAL TIME RESPONSE

Simulation tools (i.e. well calibrated hydraulic and water quality models) can be linked to SCADA real-time databases allowing for continuous, high-speed modeling of the pressure, flow, and water quality conditions throughout the water distribution network. Such models provide the operator with computed system status data within the distribution network. These "virtual sensors" complement the measured data. Anomalies between measured and modeled data are automatically observed, and computed values that exceed predetermined alarm thresholds are automatically flagged by the SCADA system. The operator, upon identification of an occurrence, can take appropriate action to either eliminate or contain the danger to public health or service interruption, or failing that, is able to map out the extent of the service disruption to guide both utility crews and emergency response units. Having taken corrective action, the operator can use the predictive modeling capability to extrapolate the future system performance. For example, in the case of introduction of contamination into the distribution system, the dispersion and dilution as a function of time can be calculated. By real time monitoring, the operator can continually update and adjust the model. Similarly, alternate water supply strategies can be quickly modeled and evaluated, guiding the operator until the situation is remedied or stabilizes.

1.2. CONTINGENCY PLANNING

The best time to prepare for an emergency, no matter how remote the possibility, is before it happens. The US Federal Government has required water suppliers to provide written emergency action plans. The utility engineer using the hydraulic modeling system has a powerful learning tool to assist in decision-making and planning. Any number of scenarios may be mapped out and the appropriate responses documented for future use. Operators can be trained to use the model to simulate various scenarios in advance of a critical need, including failure of critical facilities and introduction of toxic substances into the distribution system. They can learn to evaluate and choose an appropriate strategy for operation of pumping stations or settings of the control valves and to determine the most appropriate response to unusual operating conditions, as well as being able to predict the potential for health hazard or service disruption. Such exercises can also be used to identify and quantify critical points in the system and provide input in the planning of capital improvements.

In summary, proper design and integration of hydraulic modeling software, GIS, and the SCADA system allows the water utility to plan for and react to scenarios to hopefully assure reliability of service and water quality in the face of whatever man or nature can throw at us, or, at the minimum, identify and contain the damage and disruption of service.

2. Hydraulic and water quality analysis

Most commercial available software is based upon the industry standard EPANET hydraulic and water quality algorithm. EPANET was developed by the Water Supply and Water Resources Division (formerly the Drinking Water Research Division) of the U.S. Environmental Protection Agency's National Risk Management Research Laboratory. The hydraulic model used by EPANET is an extended period hydraulic simulator that solves the following set of equations for each storage node s (tank or reservoir) in the system:

$$\partial y_s / \partial t = Q_s / S_v \tag{1}$$

$$Q_s = \Sigma_i Q_{is} - \Sigma_j Q_{sj} \tag{2}$$

$$h_s = E_s + y_s \tag{3}$$

along with the following equations for each link (between nodes i and j) and each node k:

$$h_i - h_j = f(Q_{ij}) \tag{4}$$

$$\Sigma_i Q_{ik} - \Sigma_j Q_{kj} - Q_k = 0 \tag{5}$$

Equation (1) expresses conservation of water volume at a storage node while Equations (2) and (5) do the same for pipe junctions. Equation (4) represents the energy loss or gain due to flow within a link. For known initial storage node levels are at time zero, Equations (4) and (5) are solved for all flows q_{ij} and heads hi using Equation (3) as a boundary condition. This step is called "hydraulically balancing" the network, and is accomplished by using an iterative technique to solve the non-linear equations involved.

The method used by EPANET to solve this system of equations is known as the "gradient algorithm", Todini, E. and Pilati, S. [5], and has several attractive features. First, the system of linear equations to be solved at each iteration of the algorithm is sparse, symmetric, and positive-definite. This allows highly efficient sparse matrix techniques to be used for their solution, George-Liu [6]. Second, the method maintains flow continuity at all nodes after its first iteration. And third, it can readily handle pumps and valves without having to change the structure of the equation matrix when the status of these components changes.

After a network hydraulic solution is obtained, flow into (or out of) each storage node, q_s is found from Equation (2) and used in Equation (1) to find new storage node elevations after a time step dt. This process is then repeated for all subsequent time steps for the remainder of the simulation period.

EPANET's dynamic water quality simulator tracks the fate of a dissolved substance flowing through the network over time. It uses the flows from the hydraulic simulation to solve a conservation of mass equation for the substance within each link connecting nodes *i* and *j*:

$$\partial c_{ij} / \partial t = -(Q_{ij} / S_{ij}) (\partial c_{ij} / \partial x_{ij}) + \theta(c_{ij}) \quad (6)$$

Equation (6) must be solved with a known initial condition at time zero and the following boundary condition at the beginning of the link, i.e., at node i where $x_{ij} = 0$:

$$c_{ij}(0,t) = \frac{\Sigma_k Q_{ki} c_{ki}(L_{ki},t) + M_i}{\Sigma_k Q_{ki} + Q_{si}} \quad (7)$$

The summations are made over all links *k,i* that have flow into the head node (i) of link *i ,j*, while L_{ki} is the length of link *k,i*, Mi is the substance mass introduced by any external source at node i, and Q*si* is the source's flow rate. Observe that the boundary condition for link *i, j* depends on the end node concentrations of all links *k, i* that deliver flow to link i, j. Thus Equations (6) and (7) form a coupled set of differential/algebraic equations over all links in the network. Water quality simulator uses a Lagrangian time-based approach to track the fate of discrete parcels of water as they move along pipes and mix together at junctions between fixed-length time steps. These water quality time steps are typically much shorter than the hydraulic time step (e.g., minutes rather than hours) to accommodate the short times of travel that can occur within pipes.

By employing these features, EPANET engine can be used to study such water quality phenomena as:

- Blending water from different sources
- Age of water throughout a system
- Loss of chlorine residuals
- Growth of disinfection by-products
- Contaminant propagation events

3. The use of models

Models of water supply networks (combined with GIS and SCADA) can be used as an instrument for increasing the public safety by providing answers to questions such as:

- How can we modify the water supply system or the operational procedures in order to reduce risks?
- How should we react, if an incident occurs?
- How do we get back to normal supply, when an incident has occurred?
- How will the supply of main be cut off if contaminated?
- What amount of time is required to flush each contaminated area?
- What ncighborhoods are affected by each main?

3.1. HOW CAN WE REDUCE RISKS?

Any given tap receives water, which arrives though a number of pipes in the supply network, the transport route, and ultimately comes from a source. However, in order to achieve maximum supply security in case of pipe failures or unusual demand patterns (such as fire flows) water supply networks are generally designed as complicated, looped systems, where each tap typically can receive water from several sources and intermediate storage facilities. This means that the water from any given tap can arrive through several different routes and can be a mixture of water from several sources. The routes and sources for a given tap can vary over time, depending on the pattern of water use.

A model can show:

- Which sources (well-fields, reservoirs, and tanks) contribute to the supply of which parts of the city?
- Where does the water come from (% distribution) at any specific location in the system (any given tap or pipe).
- How long time has the water been traveling in the pipe system, before it reaches a specific location.

One way to reduce the risk – and simplify the response to incidents – is by compartmentalizing the water supply system. If each tap receives water from one and only one reservoir, then pollution of one reservoir will affect one well-defined and relatively smaller part of the city. If a toxic substance is injected into any section of the water supply system, then one and only one part of the supply system will be polluted, thus reducing the potential risk in terms of the number of people involved.

Compartmentalizing the water supply system will reduce the spreading of toxic substances. On the flip side, it may increase the concentration of the toxic substance. It is also likely to have a negative impact on the supply of water for fire flow and on the robustness of the water supply network in case of failures of pipes or other elements. These problems can be eliminated, if the compartmentalization is done properly, allowing selected valves to be opened in case of fire emergencies or pipe failures.

3.2. HOW SHOULD WE REACT, IF AN INCIDENT OCCURS?

Cities are now (if not before) establishing emergency and preparedness plans covering this kind of incidents. US Federal Law now requires the preparation of written emergency action plans for water utilities serving 3100 or more customers. A model is an invaluable tool in the preparation of such emergency plans. The model will be able to simulate a wide range of emergency scenarios, and the results can be condensed into very specific instructions for the emergency officers in charge at the time of the incident.

For instance:

A number of people in an area are reported ill with symptoms leading to suspicion that this might be caused by pollution of the drinking water. By looking up in an on-line GIS system containing pre-processed model results, the authorities have the following information readily available:

- What is the source(s) of the drinking water for the affected area?
- How long time has the water been traveling from the source(s) to the taps?
- Exactly which part of the city is receiving water from the same source(s)?
- How has the toxic substance in all likelihood been spreading? And hence, who should be warned first, and where can we expect to find most of the casualties?

This information can of course lead to actions such as sealing off the affected area from not-yet affected areas, warning of people within the affected district, starting medical treatment of people living in the affected area, setting up medical emergency centers, etc.

3.2.1. *Other pre-computed information could include*

Assuming that several reservoirs were polluted simultaneously, where should we take samples in order to quickly discover pollution spreading from other reservoirs? If such samples are negative, then those segments were most likely not polluted and people living in those areas can be told that their drinking water is safe.

Or: Where do we sample within the polluted segment in order to find out exactly where the toxic substance was in fact injected? Based on concentration patterns, it is possible to rule out some locations and point towards the likely spots where the toxic substance can have been introduced into the water supply system.

3.3. HOW DO WE GET BACK TO NORMAL SUPPLY, WHEN AN INCIDENT HAS OCCURRED?

The term Artificial Recharge (AR) covers a range of technologies that typically utilize the natural cleaning capacity of natural subsoil systems to produce drinking water from surface water. The idea is to rapidly infiltrate surface water into the aquifer thereby increasing the groundwater formation and exploration possibilities. At many plants around Europe this is done in large plants where surface water is lead to large basins where it infiltrates. Traditionally these plants are operated based on measurements of water quality on a regular basis and in many cases on real-time measurements of various flow-related parameters. This AR, or mixing together of polluted and fresh water to bring the polluted water to acceptable standards for use, is one method of dealing with an attack.

Cleaning the pipe system by flushing it is a relatively simple method, but the model is needed in order to ensure that pockets of polluted water will not remain in the system after the flushing. These reactions should be pre-planned for each segment of the pipe system, leading to instructions such as:

Flush for 3 hours by this and this method. Then change the flow in this and this manner and flush for another 2 hours."

"Turn off this main and drain via fire hydrants."

"To dilute, open this valve for water supply to this subdivision."

Etc…..

The four illustrations show the geographical extent of the pollution, at time 0, 2 hours, 4 hours and 24 hours after release of the substance in the point marked with an arrow (SOURCE TRACING). Parts of the network are not exposed to the pollution at all, as these areas are supplied from other sources.

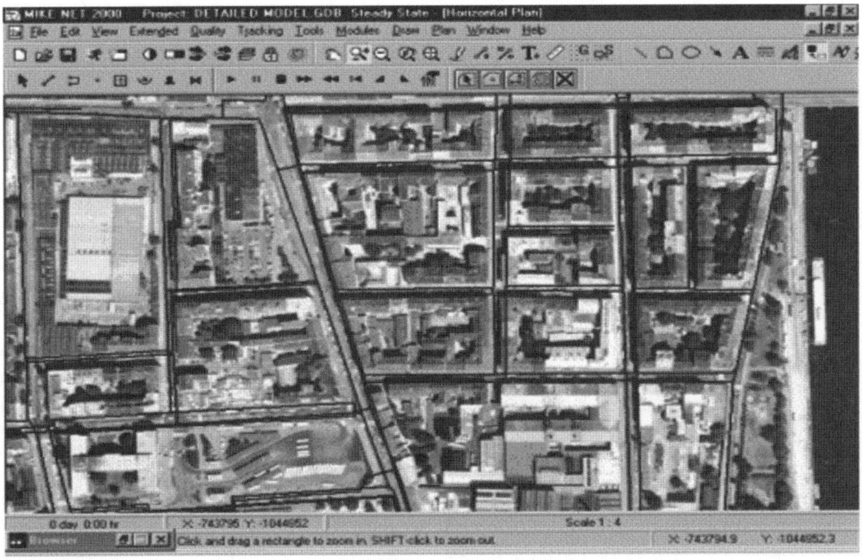

Figure 1. Spreading of pollution simulated by model, MIKE NET

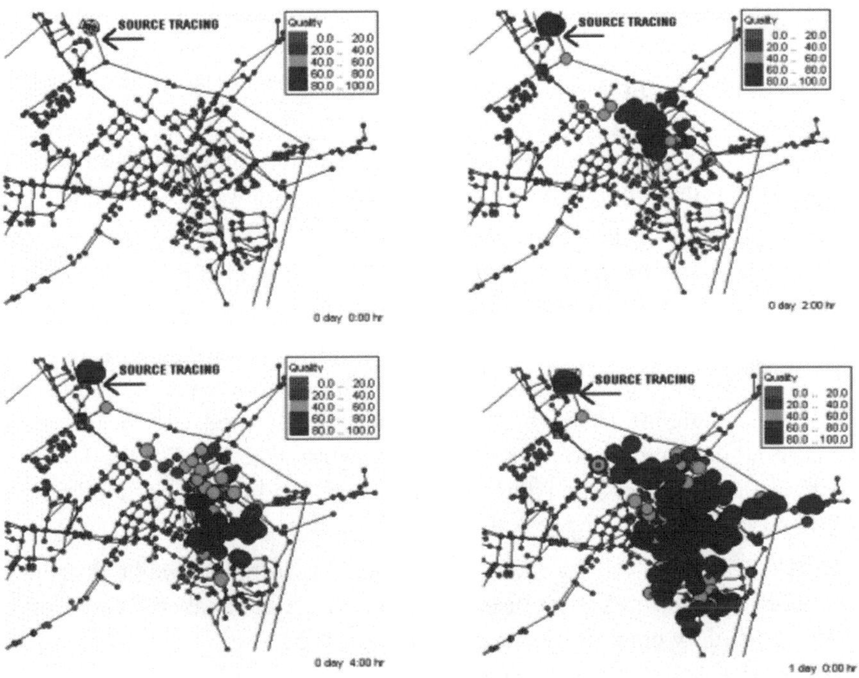

Figure 2. GIS used at a front of a model, MIKE NET Software

A GIS used as a front end for the model results will enable the emergency response staff to point-and-click on a specific address and get information about sources, areas affected, etc. The two figures below show another way of illustrating the spreading over time of a pollutant in a water supply system.

Figure 3. Spreading over time of a pollutant in a water supply system, MIKE NET Software

4. Levels of information

To control, manage, and model the water supply and water distribution networks, the following levels of information are available:

1. SCADA. The basic geometric information about the system and time varying measurements from the SCADA system. The SCADA system is used to control the physical system.
2. Strategic model. The strategic model, which is connected directly to the SCADA system and subsequently, can be set up to adapt information from the measurements. This model is continuously updated based on the measured data and it provides the hydraulic and water quality results for any part of the network. The strategic model can also be used for modeling IF-THEN scenarios and for system planning and forecasting.
3. Detailed models. All together, they cover the complete water distribution network. These models are used for detailed offline analysis, and they can reuse boundary conditions from the strategic model.

While the strategic model will be connected online the detailed models will normally be used offline. It is however very easy to generate boundary conditions for the detailed models based on actual measurements and results from the strategic model. The detailed models can therefore be used to analyze the actual situation in the system as well as to provide normal off line analysis of hydraulic, and water quality parameters.

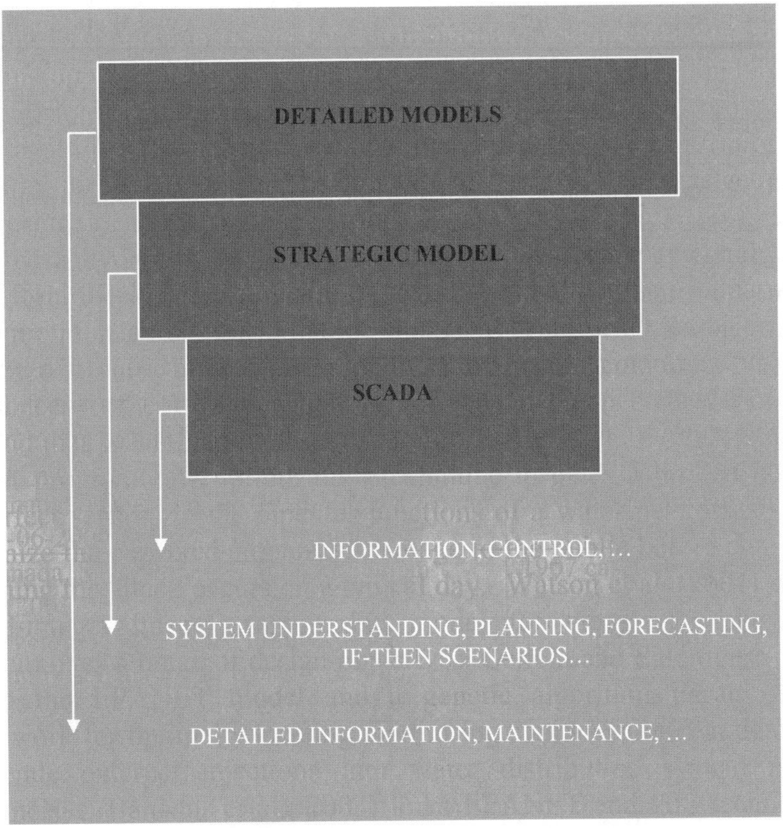

Figure 4. Levels of information

5. SCADA systems

Supervisory control and data acquisition (SCADA) is a system that allows an operator to monitor and control processes that are distributed among various remote sites. There are many processes that use SCADA systems: water distribution and treatment utilities, hydroelectric, natural gas, etc. SCADA systems allow remote sites to communicate with a control facility and provide the necessary data to control processes. SCADA is a computerized control and monitoring system. There are four major components to a SCADA system:

- Field Instrumentation
- Remote Stations

- Communications Network
- Central Monitoring Station

Field Instrumentation refers to the sensors and actuators that are directly interfaced to the plant or equipment. They generate the analogue and digital signals that will be monitored by the Remote Station. Signals are also conditioned to make sure they are compatible with the inputs/outputs of the RTU or PLC at the Remote Station.

The Remote Station is installed at the remote plant or equipment being monitored and controlled by the central host computer. This can be a Remote Terminal Unit (RTU) or a Programmable Logic Controller (PLC).

The Communications Network is the medium for transferring information from one location to another. This can be via telephone line, radio, or cable.

The Central Monitoring Station (CMS) refers to the location of the master or host computer. Several computer workstations may be configured on the CMS, if necessary. It uses a Man Machine Interface (MMI) program to monitor various types of data needed for the operation. The following is a sample configuration of a SCADA system for a water distribution system.

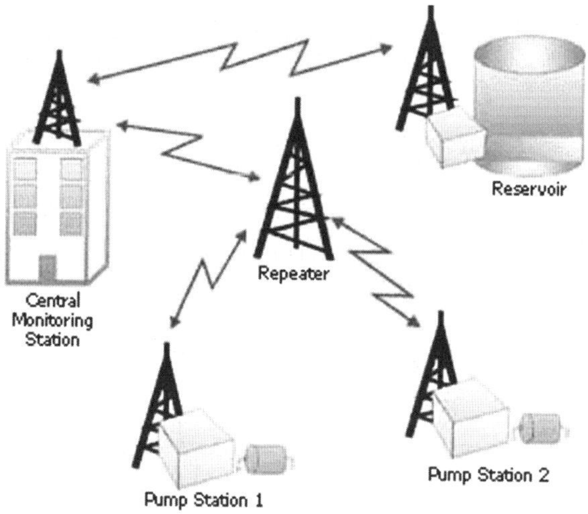

Figure 5. SCADA system for water distribution network

6. MIKE NET online

The developed linked package provides the capability to model the water distribution system in real-time, providing on-line modeling and monitoring of the system. This is essential when performing emergency response and it can greatly assist in confirming normal system performance, system trouble-shooting, improvement of system operations, and projection of the current

operating scenario. The linkage is generic so that it can be linked to any existing SCADA monitoring system.
The system consists of two modules:

- On-Line
- Off-Line

ON-LINE operates on the top of the SCADA system and it performs an on-line analysis of the system. The model results are stored back into the SCADA database; The On-Line viewer is used to display detailed model results.

OFF-LINE is used to model IF-THEN scenarios, model system breakdowns, and predict system behavior based on the demand and control rules prediction. Microsoft Access is the database used to store and maintain model alternatives.

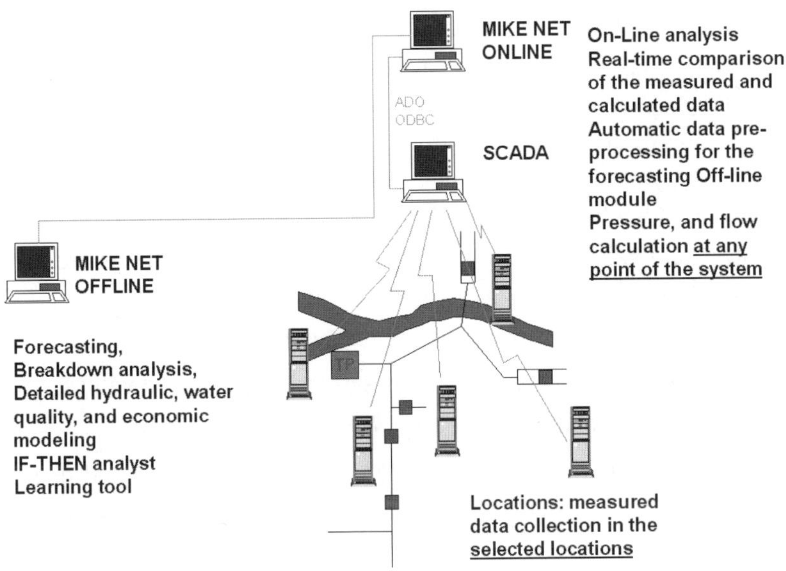

Figure 6. MIKE NET SCADA System Diagram

6.1.1. *The On-Line setup*

On-Line is developed as a service application that can be linked to any SCADA system running on Microsoft Windows NT4, Windows 2000, and Windows XP. The main advantage of this implementation is that it starts automatically with Microsoft Windows operating system as a service process; it cannot be switched on or off without the administrative right; and it is can be controlled by the SCADA monitoring system.

During each cycle, all measured SCADA data is imported into the network model and the model parameters updated. Then, a hydraulic and water quality model is automatically performed. After the analysis, output data from the model is stored in the SCADA historical database, as well as displayed on the screen. Animations of computed values, such as water quality, can be performed.

All the measurements are checked and validated with standard modules, which will flag potentially "bad" data and – if possible – fill in gaps in the time series. This means that only validated data will be transferred and used as boundary conditions in the strategic model. This is very important in order to avoid using wrong data and thus risking the use of results based on these incorrect measurements for decisions about the system. To exchange the data between On-Line and the SCADA system, ODBC, ADO, or native drivers are used.

On-Line operates in an infinite cycle of predefined time steps—such as every 15 minutes. Each cycle consists of the following steps:

1. Input data is read from the actual SCADA database, including analogue measured values of pressure, water level in storage tanks, and discharge, analogue measured water quality parameters, such as the chlorine concentration, analogue measured values of valve openings, and binary input values indicating the status of pumping stations, control valves etc.
2. Calculated demands, and calculated flow control valve settings are used to balance the input data and to overcome non-measured demands.
3. Input data is checked for errors and the gaps are filled.
4. Input file for the analysis is modified based on the measured values.
5. Hydraulic (and optional water quality) analysis is performed.
6. The output data is stored in the SCADA historical database and displayed on the screen. It is possible to store any computed parameters, such as values of pressure, water level in the storage tanks, and discharge, travel time along predefined paths, water quality parameters, such as the chlorine concentration, turbidity, water age, source tracing, etc., reservoir volume changes and residual volume, pump power costs and variable water production costs.

Continuous evaluation of the measured and computed values of pressure and flow validates the integrity of the calibrated model. Usually, validation of the model is done once a year. However, observed data are archived. Once the data accumulate, authorized personnel can activate genetic algorithms to automatically calibrate the water network.

6.1.2. *The Off-Line setup*

Off-Line enables the user to load a previously stored network model - which has been automatically prepared and analyzed by On-Line. This allows the user to inspect the stored computed model in greater detail in order to look for water distribution network problems, etc. MIKE NET Off-Line contains the data pre-processor, based on the Microsoft Access database allowing the user to control the process of modeling the selected alternatives. The key function of Off-Line module is prediction of the hydraulic, water quality, and economic parameters based on the pre-defined or forecasted behavior of the system parameters.

6.2. SMVAK, NORTHERN MORAVIAN REGIONAL WATER SUPPLY SYSTEM

The regional water supply system of Ostrava supplies with high quality water more than one million inhabitants of Northern Moravia. The system consists of more than 380km of main water supply pipelines, 60 tanks and 3 main water sources with 10 main pumping stations. The flow capacity of produced water is above 4850l/s and delivered to more than 100 towns within the region. The mathematical model of the whole water supply system was created, calibrated and verified for both steady state and extended period simulations.

The On-Line computational module is installed on the central control system server that receives data from more 41 local monitoring stations. These monitoring stations measure data on flow, pressure and water elevation at different intervals. However, this data is reorganized for a time step of 15 minutes. Sometimes this lack of synchronization in measuring data at various monitoring stations can result in delays in data transfer for analytical purposes.

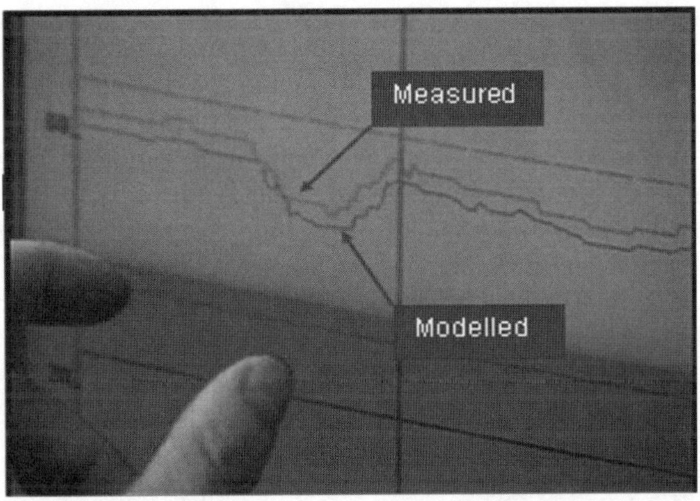

Figure 7. Comparison of measured and modeled data is on-line monitored by the SCADA system

The On-Line module is implemented within the regional water distribution system and it serves as an automatic super-sensor that instantly evaluates the observed data and synchronizes the model for quick decision-making. It is also used by the network operators as a learning and contingency planning tool to model any network situations that may occur such as pipe breakdowns, water supply failure, pump trip-off and similar. The model was implemented as it is described in the section 4.1.

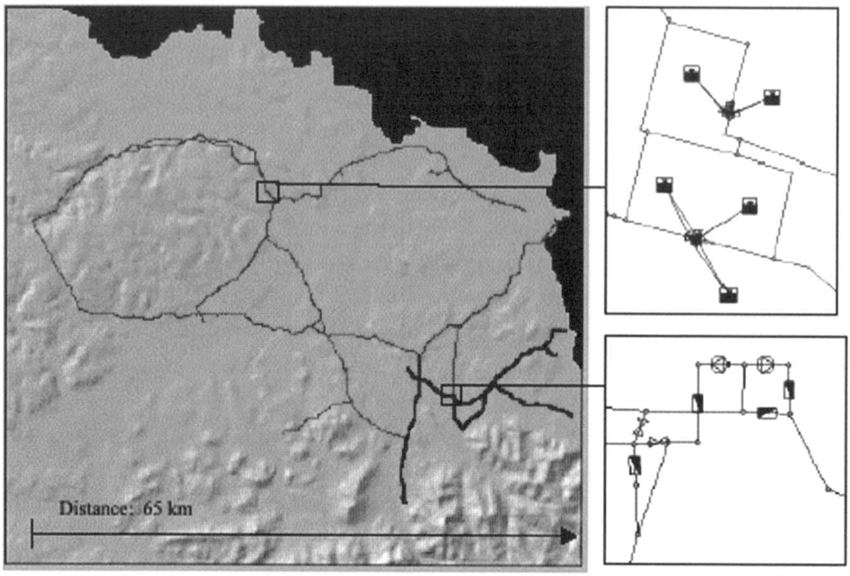

Figure 8. The pilot project area (November 1998) and the current project area (September 1999)

The On-Line module was developed as a supervisory and supportive tool for the network operators allowing them to quickly evaluate and predict the system behavior and use the modeled (forecasted) results in their decision making. The calibration level of the hydraulic model was significantly improved; by running the model in the on-line mode for more than 3 years it was possible to achieve more than 15% accuracy during the extended period simulations for almost any system operation. The expected benefits were: replace some of the physical sensors by the virtual (modeled) sensors, predict pollution tracing, train new operators using the model scenarios, and increase the water availability under unexpected incidents. The proposed solution proved to be a suitable tool for such requirements.

Reference: Severomoravske Vodovody a Kanalizace a.s., 28 rijna, 709 45, Ostrava, Czech Republic, 1998-1999, Scada: Retos_NT, FCC Folprecht – Logica

6.3. SCVK, NORTHERN BOHEMIAN REGIONAL WATER SUPPLY SYSTEM

The regional water supply system of Teplice supplies with water more than 1.1 million inhabitants of Northern Bohemia and there are 5 regional control rooms controlling the whole water supply system. The whole system consists of 47 water treatment plants, 948 storage tanks, and 568 pumping stations on the area of 8 700 km2. MIKE NET On-Line computational module is installed on the central control system server of the node Chomutov, which consists of more than 200km of main water supply pipelines, 26 tanks, 3 main water sources, 2 main pumping stations, and 60 control valves. The flow capacity of produced water is above 900l/s. The mathematical model of the water supply system was calibrated and verified for both steady state and extended period simulations. To incorporate the modeling tool within the SCADA system, more than 200 nodes and links are used to exchange the data including:

- 97 input parameters from SCADA, 67 node input parameters, 30 link input parameters
- 115 output values from the model, 61 derived parameters

Two specific issues had to be addressed during this project: manually controlled valves and calculated demands. Valves, which can be manually open or closed, were entered into a fictive location of the SCADA systems and their state is continuously set by the operators whenever these valves are manually controlled or the system is parameterized. Calculated demands are demands, where the mass balance is not fulfilled by the measured data and the mass difference has to be derived based on the measured data.

The On-Line module is implemented within the regional water distribution system to evaluate the difference between the observed and modeled data and to automatically prepare the ready-to-use model for quick decision-making by the Off-line forecasting module. This forecasting and planning module can model the system behavior in the next time frame such as 24, 48, 72 hours based on the history of the measured data. The model was implemented as it is described in the section 4.1.

The main objective in developing the On-Line model was to decrease the operational costs of the system by optimizing the pumping station operation, improving control rules used to control the inflow into the storage tanks, and decreasing the time necessary to model required alternatives. These plans turned to be too ambitious to be achieved within the short period of time, during which the pilot model was developed. The main problem was to establish formal procedures at the contractor side to devote a person (modeler) who could use the model on the daily basis. It was necessary to reorganize the modeling and operational team and establish closer links between GIS, operational, and maintenance divisions within the organization. The remaining parts of the system are being developed jointly by the contractor and consultants.

At present, the system managers use the model results to evaluate the impact of the network operation on the variable costs automatically calculated by the On-Line module.

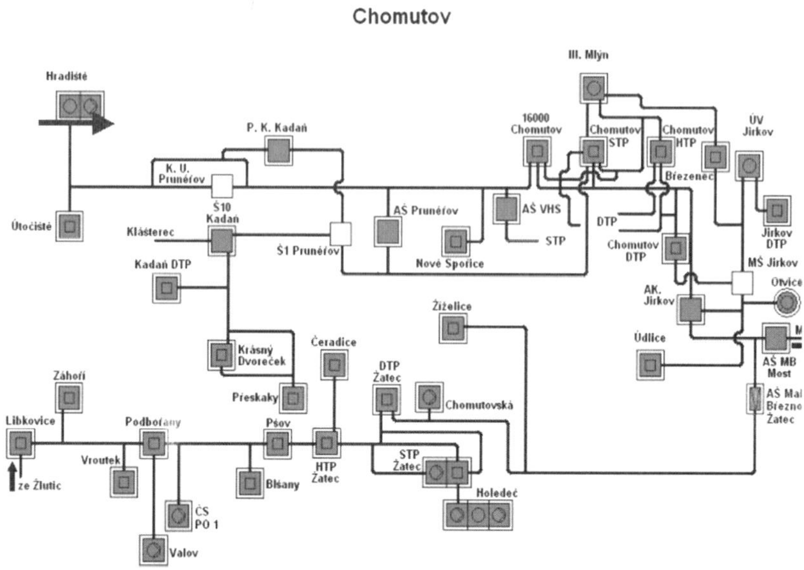

Figure 9. Scheme of the SCVK Regional Water Supply Scheme

Reference: Severoceske Vodovody a Kanalizace a.s., Teplice, Czech Republic, 2001-2002, Scada: Retos_NT, FCC Folprecht – Logica

6.4. OSLO, DETAILED MODEL OF THE WATER DISTRIBUTION SYSTEM

Water supply and distribution network of Oslo Kommune consists of 50 distribution zones with more than 25,000 pipes; it supplies water to 530,000 inhabitants of the City of Oslo and the municipality of Ski. The leakage is app. 35%. The GIS database model of the network includes all network elements required for modeling. Data is stored in Arc Info system; GEMINI (Arc Info extension) provides operational data in SCADA system and the detailed 1:1 data link to the model. The detailed model of the whole system is developed and is used for the system optimization and the water quality modeling.

In order to link iFIX SCADA database with the model, the data dictionary file is used to define the data, which is transferred between iFIX Dynamics Real Time Database and MIKE NET On-Line database.

The On-Line module is implemented within the detailed model of the selected distribution zone to evaluate the difference between the observed and modeled data and to develop an online calibrated model. Based on the modeling results and experience, the on-line model will be extended on the whole model of Oslo water distribution network. The on-line model is developed as a part of the systematic approach to modeling, where both the detailed and trunk models are developed using a real-time GIS link.

All commercial and industrial services are metered. Private households can choose between metered services or fixed services based on an estimated consumption of 1.3 m3 per m2 of living area; so far only about 220 private household customers have chosen the latter. Full functionality to carry out a water balance is therefore required for On-Line simulation. The application balances the quantity of water entering the system and that leaving the system through time by adjusting un-accounted for water (UFW) demand including a water loss (consumption) component assigned on the basis of background leakage. Actual network demand is automatically redistributed to each node in the network in order to match the zone inflow obtained from the SCADA measurements; method of two coefficients is used to calculate the appropriate junction demands.

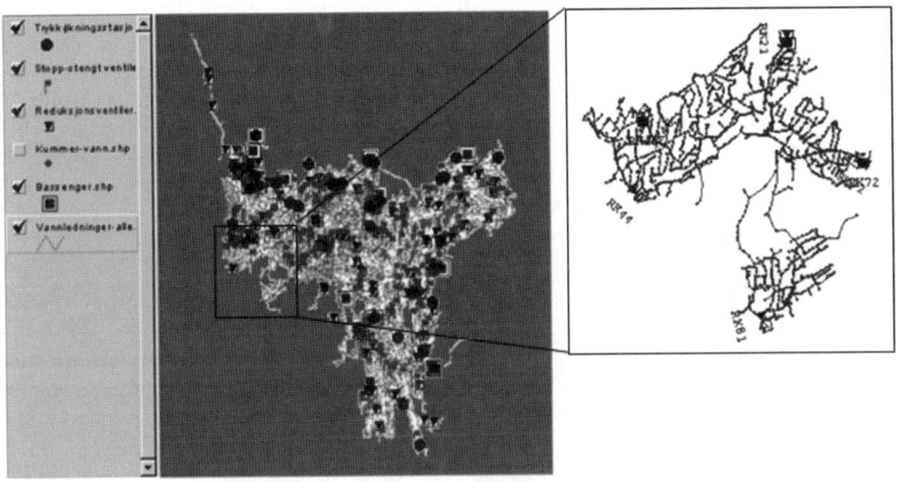

Figure 10. GIS Model of Oslo Water Distribution System

Reference: Oslo Kommune, Herslebs gt. 5, Oslo, Norway, 2001-2002 (in progress), Scada: Intellution iFIX, Intellution

7. Conclusions

Linking calibrated hydraulic and water quality model to SCADA real-time databases allows for continuous, high-speed modeling of the pressure, flow, and water quality conditions throughout the water distribution network. Such model provides the operator with computed system status data within the distribution network. These "virtual sensors" complement the measured data. Anomalies between measured and modeled data are automatically observed, and computed values that exceed predetermined alarm thresholds are automatically flagged by the SCADA system. The operator, upon identifying an occurrence, can take appropriate action to either eliminate or contain the danger to public health or service interruption, or failing that, is able to map out the extent of the service disruption to guide both utility crews and emergency response units. Having taken corrective action, the operator can use the predictive modeling capability to extrapolate the future system performance. For example, in the case of introduction of contamination into the distribution system, the dispersion and dilution as a function of time can be calculated. By real time monitoring, the operator can continually update and adjust the model. Similarly, alternate water supply strategies can be quickly modeled and evaluated, guiding the operator until the situation is remedied or stabilizes.

References

Brosnan Thomas M. Early Warning Monitoring to Detect Hazardous Events in Water Supplies, An ILSI Risk Science Institute Workshop Report, ISBN: 1-57881-075-2, 1999.

Bunn, S, Helms, S, "Application of an expert system to control treated water distribution". Vodafone House, Auckland, New Zealand, sbunn@beca.co.nz, New Plymouth District Council, New Plymouth, helms@npdc.govt.nz, 2001.

Cameron, R.W, Barret, R.J., Cazottes, N, Jarrige, P.A, Tocqueville, L, "Link scada with network analysis system: what, why and how?", Proceedings of the Third International Conference on Hydroinformatics, ISBN 90 5410 983 1, Copenhagen, Denmark, 1998.

Hosner, Marsha Use of a Model to Help Secure a Water System, DHI North America Conference, Orlando, June, 2002.

Haimes, Y.Y.; Matalas, N.C.; Lambert, J.H. ; Jackson, B.A. ;& JRF Fellows. Reducing Vulnerability of Water Supply Systems, 1998.

Ingeduld, P; Turton, G. OnLine Analysis of Water Supply and Water Distribution Systems, NZWWA Conference, Auckland, New Zealand, 2002.

EPANET Methodology, Water Supply and Water Resources Division of the U.S. Environmental Protection Agency's National Risk Management Research Laboratory, 1999.

MIKE NET User Guide, DHI Water and Environment, Horsholm, Denmark, 2001.

USE OF UV-VIS SPECTROMETRY FOR ALARM PARAMETERS IN DRINKING WATER SUPPLY

R. PERFLER*, G. LANGERGRABER, W. LETTL AND N. FLEISCHMANN

Institute for Sanitary Engineering and Water Pollution Control, BOKU - University of Natural Resources and Applied Life Sciences, Vienna, Muthgasse 18, A-1190 Vienna, Austria.

s::can Messtechnik GmbH, Brigittagasse 22-24, A-1200 Vienna, Austria

hydrophil, Lerchenfeldergürtel 43, Top 6/3, A-1160 Vienna, Austria

Abstract. Alarm parameters are in many ways different from measurements of well-defined chemical substances. Being confronted with an increasing number of potentially harmful compounds as well as financial and logistical constraints, new variables (such as alarm parameters) that allow for an integrated assessment or for a first screening can be a solution. To monitor for surrogate or aggregate variables can be a useful strategy to overcome some of the constraints. It must be conceived that this can go along with losses in terms of comparability of results and even in tailor-made variables. Spectral data and their evolution over time are rich in information and compensate for losses due to aggregation and generalisation. Therefore it can be expected that alarm parameters developed from spectral data are transferable, accurate and selective to an extent that is beyond the state-of-the-art. Examples for application at waterworks show the actual practical use of *time-resolved delta spectrometry* and generated alarm parameters from spectral data.

Keywords: Alarm parameters; early warning; on-line monitoring; time-resolved delta spectrometry.

To whom correspondence should be addressed: Reinhard Perfler, Institute for Sanitary Engineering and Water Pollution Control, BOKU - University of Natural Resources and Applied Life Sciences, Vienna, Muthgasse 18, A-1190 Vienna, Austria. reinhard.perfler@boku.ac.at

J. Pollert and B. Dedus (eds.), Security of Water Supply Systems: From Source to Tap, 85–98.

1. Introduction

The goal of alarm systems (early warning monitoring systems) is to reliably identify low probability/high impact contamination events (chemical, microbial, radioactive) in source water or distribution systems in a timely manner. These events may result from

- extreme natural events that might compromise water quality (e.g. flooding, turbidity, anoxia, algal blooms),
- extreme anthropogenic events (e.g. inadvertent discharges/spills) or intentional discharges/actions (e.g. vandalism, terrorism), and
- discharges of compounds that might pose chronic health risks (e.g. xenoestrogens, biocides, pharmaceuticals, pesticides).

One general solution for monitoring would be to use indicators like microbial cultures or caged organisms. However, conventional biological systems still suffer from a lack of selectivity and validation, and there are not many reports available about successful applications using such systems.

Another solution is based on indirect measurement, and on the assumption that even a small contamination can be detected as a deviation from a reference signal. A broad-band picture of the overall water quality is monitored with the help of a combination of physical and chemical sensors. Any deviation of a pre-defined reference condition is used as a warning signal. The reference or baseline is normally generated from historical samples, so the system must be "trained". This approach is used by time-resolved delta spectrometry.

2. Alarm parameters

2.1. REQUIREMENTS FOR ALARM SYSTEMS/PARAMETERS

The typical function of alarm systems is to separate reference situations (e.g. *background concentrations* and "*normal*" *variations*) and “alarm” variations for the parameters to be used.

According to Brosnan (1999) the following requirements for alarm systems and parameters can be defined:

- quick response to suspect quality changes (results in the need for real-time measurement),
- high sensitivity and at the same time low probability for false alarms,
- clear interpretability of signals and alarms,
- broad-band response to diverse contamination sources (at the cost of low selectivity),

- reliability,
- reproducibility,
- remote accessibility,
- low costs of purchase and operation to promote a widespread application,
- high security access, including secure data exchange, high security reserves and redundancies on all levels.

Less important for alarm systems/parameters are:

- parameter selectivity, and
- compliance to analytical laboratory norms

Broadband optical instruments like spectrometers fulfil most of the requirements listed above. Spectrometers, for example, need little maintenance since they only have glass windows in contact with the water, and prove to be highly robust and have long-term stability even under harsh conditions (e.g. Hofstaedter et al., 2003). Zero-compensated systems like 2-beam-designs are distinctly superior to non-compensated systems. With a 2-beam-design it is possible to increase the long-term stability of the spectrometer due to, for example, compensating altering effects and detecting instrumental failures.

2.2. ALARM PARAMETERS DERIVED FROM SPECTRAL MEASUREMENTS

2.2.1. *Direct and selective spectrometric measurements*

On-line spectrometry can replace several complicated and expensive on-line analysers at the same time. The identification of single substances or substance groups is limited to those substances that

7. are detectable in the UV/VIS spectrum and
8. are implemented in the algorithm evaluating the spectra.

Parameters resulting from spectral information include surrogate parameters (e.g. turbidity, organic matter and colour), and single substances (e.g. nitrate, nitrite, phenol, BTEX, nitrobenzene).

Changes in the low ppb range can be monitored using on-line spectrometry. However, the selectivity is rather low and single substance identification below 100 ppb is difficult if not impossible for many substances. UV/VIS spectrometry is typically not selective enough to measure micro-pollutants at the low concentration levels that are defined in most drinking water norms based on the WHO risk levels of 10^{-6} (WHO, 1996). Thus, the advantages of online spectrometry are more evident in low probability/high impact concentration ranges than in detection of long-term chronic risks.

2.2.2. *Monitoring of indicator parameters instead of contaminants*

Indicator parameters can be used instead of monitoring contaminants directly. For example, not all **Polycyclic Aromatic Hydrocarbons** (PAH's) are usually monitored but instead, only six indicator substances. Another example would be to use nitrate measurements as an indicator for pesticides in groundwater.

2.2.3. *Monitoring of matrix parameters instead of contaminants*

The whole UV/VIS spectra can be used as a fingerprint of the water composition ("matrix"). For monitoring of matrix changes the spectra have to be evaluated. Very small changes can be detected by tracing differences between spectra over time or space. Sometimes the detected changes are not directly related to known substances, but nevertheless can provide a sensitive alarm parameter.

There are several ways to evaluate the provided spectral information:

- qualitative interpretation of spectral deviations from a site specific reference spectrum (e.g. peaks, shoulders, gradients, analysis of the derivative spectra, etc.),
- comparison of spectral differences between measuring points in a measurement network, and
- evaluation of changes of the spectral features over time (anthropogenic changes are typically faster than natural changes).

As for all alarm systems, a well-defined approach to guarantee data quality and data interpretation is needed. Specific baselines must be evaluated and deviations that lead to an alarm must be defined. It is always a difficult task to distinguish between "normal" baseline variations originating from natural occurrences and the frequent small peaks that originate from other influences or impacts when using unselective and low specific surrogate parameters like turbidity. Because of the multi-dimensional information provided by UV/VIS spectrometry, it offers a much higher information potential on the "normal" baseline compared to single signals like turbidity or DOC. The combination of several sensors to a sensor array further reduces the probability of false alarms.

2.3. DELTA SPECTROMETRY

Differences of spectra can be calculated between different measurement points and over time resulting in spatially-resolved and time-resolved delta spectrometry respectively.

2.3.1. *Spatially-resolved delta spectrometry*

A "reference" is measured at a point where unimpaired and/or constant water quality can be guaranteed. At strategically chosen measurement points in the monitoring network (spectral) water quality is measured. These strategically chosen measurement points can be located e.g. along a river or after different stages of a water treatment plant. The measured data (spectra) are sent to a central database and evaluated in real-time. Using spatially-resolved delta spectrometry it is possible, for example, to distinguish between natural and anthropogenic causes of spectral deviations in rivers, or to monitor and control, and therefore optimise treatment processes. Figure 1 shows a schematic sketch of a monitoring network where the data are collected at 4 monitoring stations in a watershed and transferred to a main station where the data are evaluated.

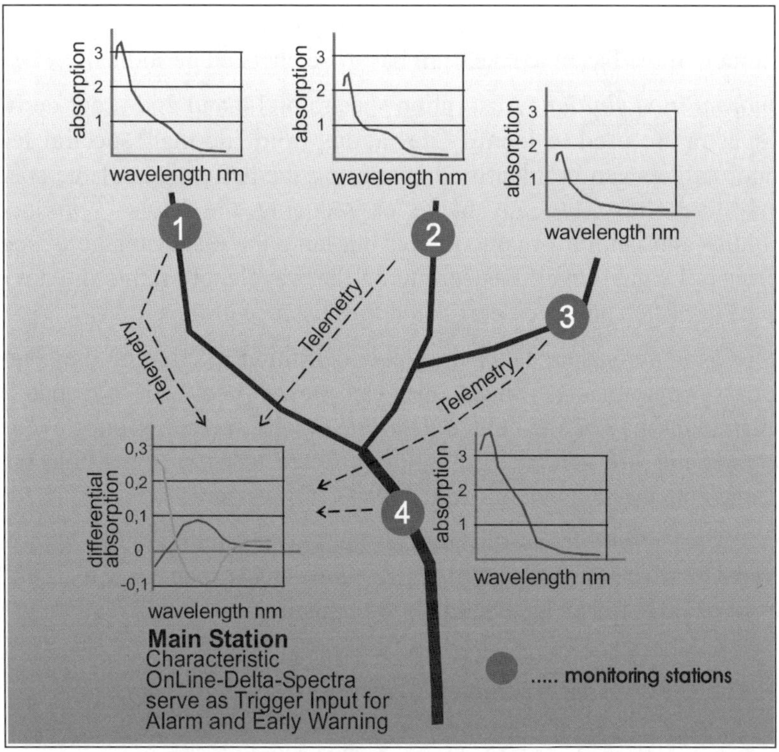

Figure 1. Schematic sketch of a monitoring network for spatially-resolved delta spectrometry (Weingartner and Ružička, 2003)

2.3.2. *Time-resolved delta spectrometry*

Also in cases where no water body can serve as a reference for unimpaired water quality it is generally possible to identify extraordinary water compositions by UV/VIS spectrometry. On the one hand, the UV/VIS spectra of natural water bodies do not provide very distinct features, like peaks and valleys. On the other hand, every water body has his own fingerprint or "baseline spectrum". This of course has to be considered in the range of natural deviation.

The definition of alarm parameters using time-resolved delta spectrometry has to involve the following periods:

- *Learning period*: During the learning period (usually some months) spectra are measured. After this period, the shape and features of the baseline spectrum is evaluated. The baseline spectrum will serve as a reference – or baseline – for later identification of extraordinary composition. The accuracy of the baseline spectrum has to be checked periodically.
- *Abnormality definition*: Absorption spectra or 1st and 2nd order derivatives of spectra are used to identify deviations from "normal" spectral features. Abnormalities can be identified best using the derivatives of the spectrum. Derivatives have the advantage of reducing the "noise" produced by turbidity and natural organic matter but have the disadvantage of reducing the overall signal level. Depending on the wavelength range the deviations can be qualified and even correlated to substance groups.
- *Alarm level definition*: The alarm level definition is based on the concept of virtual contaminants. Since one can never know in advance which contaminant may be the next to enter the system, several groups of "virtual" contaminants are generated that are expected to cover the whole range of organic contaminants visible in the UV/VIS spectrum.
- *Sensitivity definition*: Sensitivity can be adjusted individually with respect to risks involved and acceptable false alarm levels. The chosen alarm levels are based on empirical and statistical evaluation.

3. The UV/VIS spectrometer

The tested submersible UV/VIS spectrometer (spectroanalyser) measures absorbance of ultraviolet and visible light from 200 to 750 nm. The instrument is built as a compact submersible sensor enabling measurement of UV/VIS spectra with laboratory quality directly in liquid media. A single evaluation of the entire spectrum typically takes 15 seconds. Sensitivity can be adapted to the application demands by selecting the optical path length within a range of 1 - 100 mm. This opens a wide range of applications from ultra pure waters (DOC

> 10 μg/l) up to concentrated industrial wastewaters with a COD of several 1000 mg/l. The spectrometer is equipped with an auto-cleaning system using pressurized air, which proved to work extremely reliably (e.g. Hofstaedter et al., 2003; Langergraber et al., 2003b).

The sensor can be calibrated to all absorbing substances by correlating the measured spectra to concentrations. Typical applications in the water sector are organic matter (e.g. total COD, COD fractions, BOD_5, TOC, or DOC), TSS, turbidity and nitrate. For typical waters (e.g. municipal wastewater – raw and treated, river water, drinking water) a so-called "global calibration" is available as a default configuration of the spectrometer. Usually high precision can be achieved using these standard parameter sets. A second calibration step (local calibration) improves trueness, precision and long term stability of the results. A detailed description of the calibration procedure implemented is given in Langergraber et al. (2003a).

The alarm software that was used includes 8 virtual organic contaminant fractions based on derivative spectrometry. The method is protected and details can be discussed with the corresponding author.

4. Results and discussion

The example shown here describes the application of time-resolved delta spectrometry for a water body heavily polluted with hydrocarbons.

4.1. STEP 1: LEARNING PERIOD

Figure 2 shows the 3-D spectral picture of the water body over a period of 1 day. Even without sophisticated mathematical tools, peaks, valleys, and shoulders, originating from abnormal emissions to the water body, can be clearly identified. In this case a period of 3 months was needed for learning due to the strong variations that occurred. The periods between the visible peaks have been utilised for the evaluation of the baseline spectrum.

4.2. STEP 2: ABNORMALITY IDENTIFICATION

It could be shown that most of the natural deviations of the baseline spectrum can be eliminated by derivative spectrometry and therefore abnormal spectral features can be identified clearly. Figure 3 gives the 1st derivatives of the spectra shown in Figure 2. By eliminating the background noise the features like peaks and valleys become more clearly visible. Peaks originated from hydrocarbons and colour respectively can be distinguished in Figure 3.

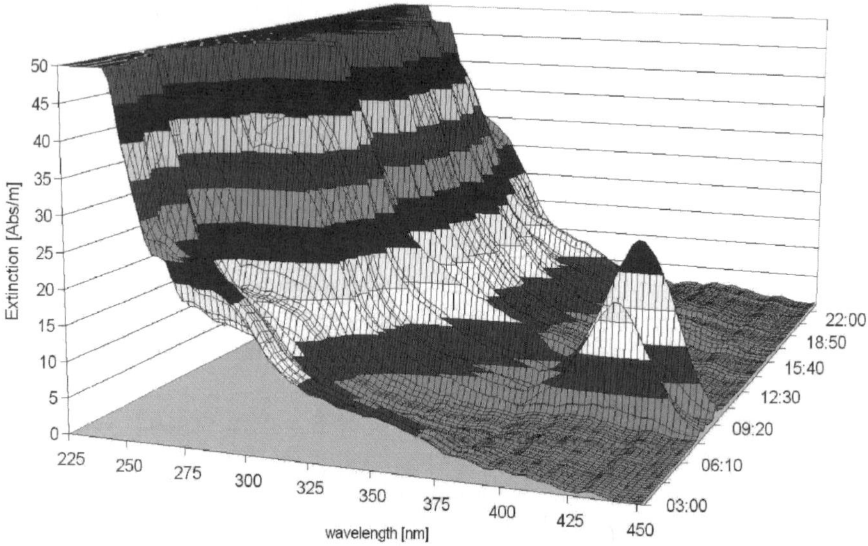

Figure 2. 3D-UV/VIS spectra plotted against time

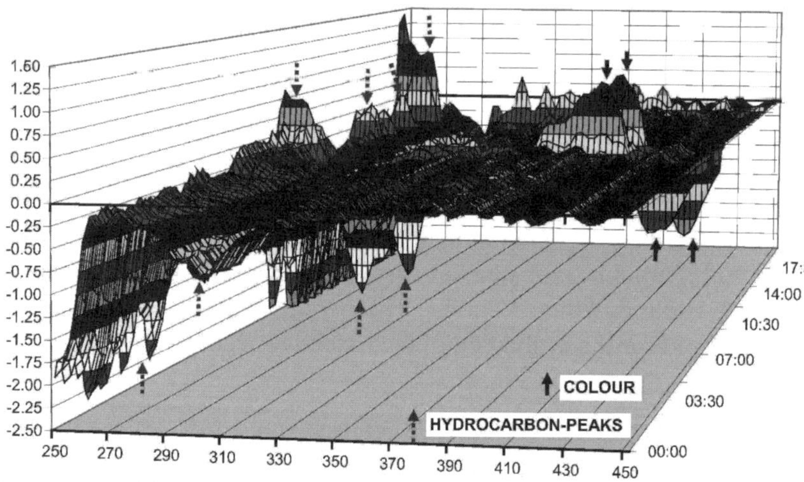

Figure 3. 1st derivative UV spectra of Fig. 2 plotted against time

4.3. STEP 3 AND 4: ALARM LEVEL AND SENSITIVITY DEFINITION

The concept of virtual contaminants was used for the definition of the alarm parameters. Eight groups of virtual contaminants have been generated. They are expected to cover the whole range of possible organic contaminants visible in the UV/VIS range.

The alarm levels are based on empirical evaluation. The normalised spectral derivatives representing the virtual contaminants are plotted against occurrence probabilities. The experience showed that waters that do not suffer severely from anthropogenic impact show a very homogeneous and uniform spectral behaviour. The probability plots are narrow banded and thus allow the setup of quite distinct and reproducible levels for most of the eight virtual contaminant groups. Thus, extraordinary emissions can be identified at satisfying probability levels.

Figure 4 shows the occurrence probabilities and the defined alarm levels for three virtual contaminants (Parameter 1 and 2, and colour). For example, for the parameter "Colour" the first and second alarm levels are reached with an occurrence probability of 99.3 % and > 99.9 % respectively.

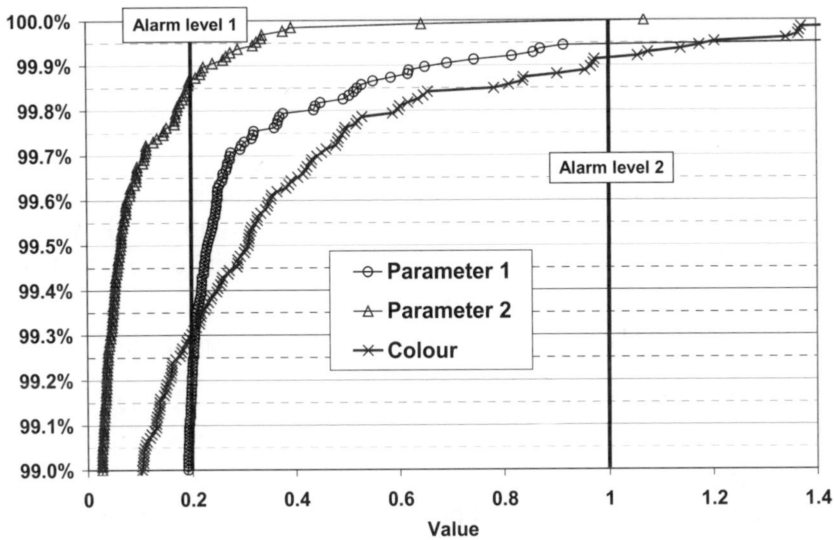

Figure 4. Definition of "Alarm level 1" and "Alarm level 2" for 3 alarm parameters

4.4. APPLICATION FOR QUALITY MANAGEMENT OF SPRINGWATER

Figure 5 shows the installation of an early warning system at the waterworks of Vienna. The application of UV-VIS was implemented for the water quality management of karstic springs at the 1st spring water mains. The UV-VIS probe is used to define extraordinarily high turbidity and SAC directly at the spring to divert turbid water before entering the mains (Zerobin, 2005).

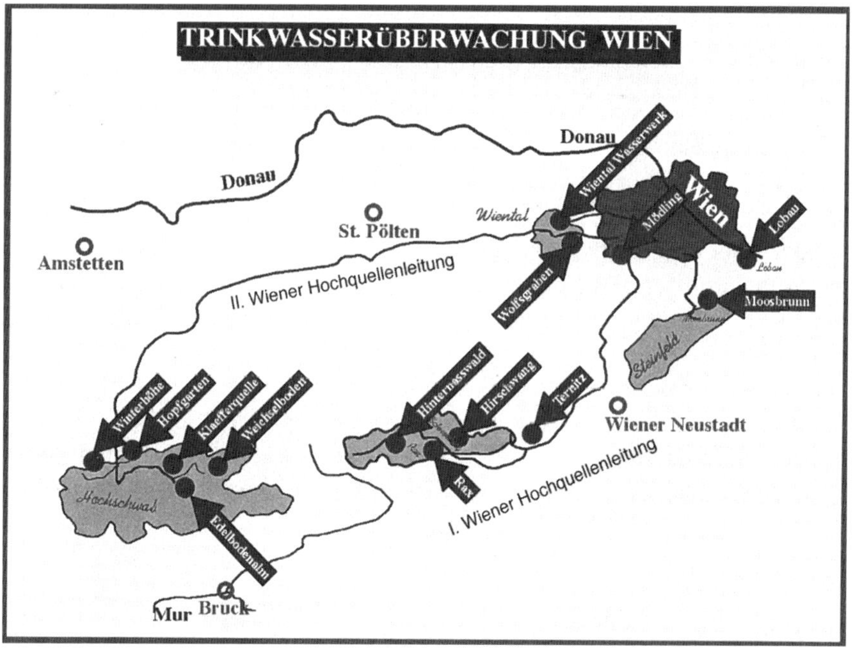

Figure 5. Online monitoring of quality parameters at watersheds of Vienna waterworks

Figure 6 shows the development of SAC and turbidity at a karstic spring during a stormwater event in the watershed. Eventually the turbid water from selected wells is diverted to manage an overall high quality of collected water

10.5. APPLICATION AT A SURFACE WATER TREATMENT PLANT

Possible applications of spectral measurements in water works include:

- Measurement of microbial contamination: One of the challenges for drinking water supply is to guarantee that after disinfection there is no re-contamination in the distribution networks. There is a strong correlation

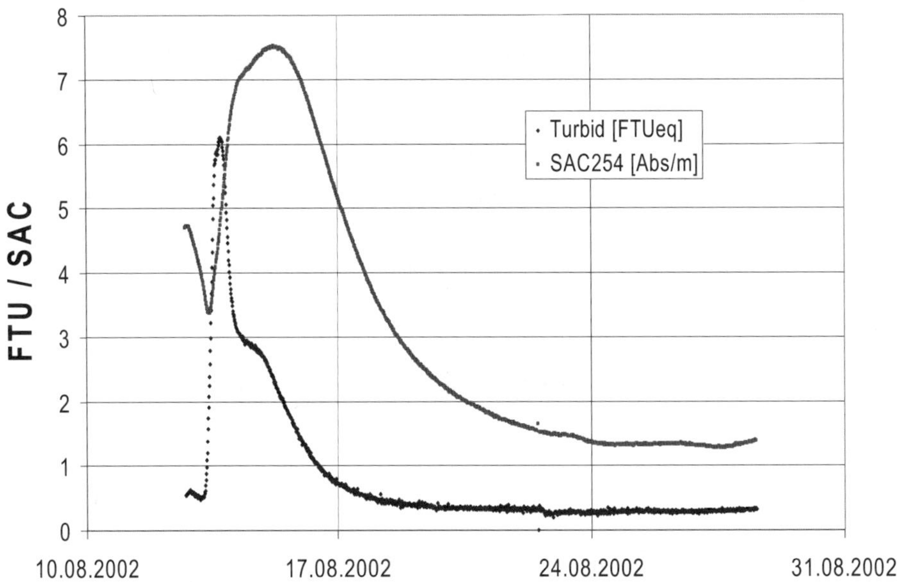

Figure 6. Development of SAC and turbidity at a karstic spring during stormwater event

between measured microbial contamination and organic matter concentration. Therefore TOC (that can be derived from spectral information) can be used as a parameter for the microbial contamination and as valuable alarm parameter for drinking water distribution nets.

- Measurement of organic carbon species: Organic carbon species change during water treatment. The spectral properties can be used to differentiate organic carbon types (e.g. degradable, absorbable, crackable by ozone, and NOM - natural organic matter), which give information for optimising the operational management of drinking water treatment (e.g. flocculation).
- Sampling trigger: Alarm parameters for event triggering can be derived using time-resolved delta spectrometry. Low-level alarms could trigger an automated sampling whereas high alarm levels could have an impact on the operation of process itself.

Within the WATERMONITOR project (Langergraber et al., 2005) the water treatment plant of Boara Polesine (I) was used as a testing site. The Boara Polesine treatment plant (Figure 7) takes about 240 l/s from the Adige River and about 80 l/s from bank side wells.

River water is treated as follows: Pre-treatment, flocculation, sedimentation (3 sedimentation tanks in parallel), and filtration (6 sand and 12 active carbon filters). Well water treatment includes oxidation and aerated grit chamber, and filtration (2 sand and 2 active carbon filters). For monitoring, the following six sampling points were available: P1 - River water, P2 - after flocculation, P3 - after filtration, P4 - Well water, P5 - Well water treated, and P6 - Common effluent.

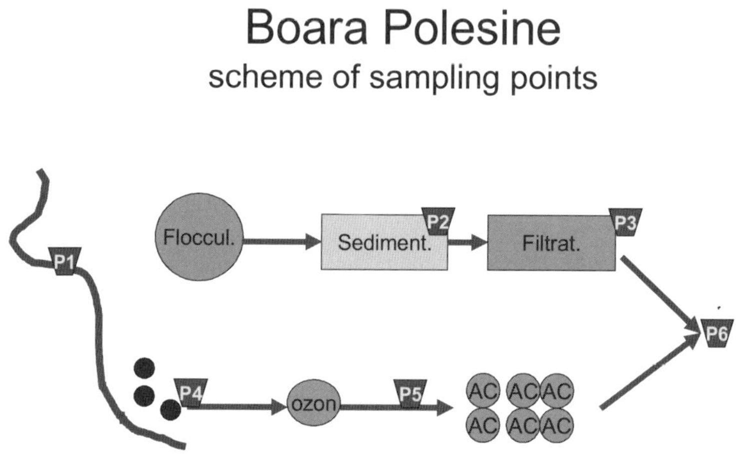

Figure 7. Development of SAC and turbidity at a karstic spring during stormwater event

The behaviour of the treatment plants could be described as well as changes in water quality during clariflocculation. Due to the strong seasonal behaviour of river water quality less data were measured to allow a final alarm parameter definition. The experience showed that a high quality of the measured spectra is indispensable when applying time-resolved delta spectrometry. Further on for monitoring of river water, only the installation of the probe directly in the river can be used.

5. Summary and conclusions

Time-resolved delta spectrometry has been applied for the definition of alarm parameters from spectral data. The "fingerprint" of a water body, the baseline spectrum, is compared with its evolution over time. The procedure for the definition of the alarm parameters includes a learning period for the definition of the reference or baseline (the normal situation), the abnormality definition, and the definition of the alarm levels and sensitivities. Derivative spectrometry eliminates most of the natural deviations of the baseline spectrum and therefore enables a better identification of abnormal spectral features. The definition of the alarm levels is based on empirical and statistical evaluation. The sensitivity is adjusted with respect to risks involved and acceptable false alarm probabilities.

The method is readily available and can be applied from wastewater (detection of abnormal emissions) to drinking water (early warning for accidents or intentional contamination). Currently the method is also tested by several waterworks. Being a qualitative or semi-quantitative method, it goes one step beyond classical parameter monitoring and may open the door to a new generation of intelligent sensors and monitoring systems. However, the success of this and other related new methods will greatly depend on the acceptance by the water community, water administration, and water legislation.

Acknowledgement

The work described is carried out within the EU-CRAFT project "*WATER management system based on innovative MONITORing equipment and DSS*" (contract no. EVK1-CT-2002-30022; duration: 01/2003-12/2004). The authors are grateful for financial support.

References

Brosnan, T.M. (ed., 1999): Early warning monitoring to detect hazardous events in water supplies. *ILSI Risk Science Institute Workshop Report*, International Life Sciences Institute, Washington D.C., USA.

Hofstaedter, F., Ertl, T., Langergraber, G., Lettl, W., Weingartner, A. (2003): On-line nitrate monitoring in sewers using UV/VIS spectroscopy. In: Wanner, J., Sykora, V. (eds.): *Proceedings of the 5th International Conference of AČE ČR "Odpadni vody - Wastewater 2003"*, 13-15 May 2003, Olomouc, Czech Republic, pp.341-344.

Langergraber, G., Fleischmann, N., Hofstaedter, F. (2003a): A multivariate calibration procedure for UV/VIS spectrometric quantification of organic matter and nitrate in wastewater. *Wat.Sci.Tech.* **47**(2), 63-71.

Langergraber, G., Fleischmann, N., Hofstaedter, F., Weingartner, A., Lettl, W. (2003b): Detection of (unusual) changes in wastewater composition using UV/VIS spectroscopy. In: Ruzickova, I., Wanner, J. (eds.): *Proceedings of the 9th IWA conference on "Design, Operation and Costs of Large Wastewater Treatment Plants" – Poster papers*, 1-4 September 2003, Prague, Czech Republic, pp.135-138.

Langergraber G., Fleischmann, N., Pressl, A., Tassinato, G., Piazzola, E., Lettl, W. (2005): Spectral data for monitoring and control of a surface water treatment plant. Oral presentation at ICA Seoul 2005.

Weingartner, A., Ružička, M. (2003): Sensors for Water Security and Alarm Systems. In: *Proceedings of the conference "Dodávka vody v krízových situáciách"*, 11 September 2003, Bratislava, Slovak Republic.

WHO (1996): *Guidelines for drinking water quality.* 2nd Edition, World Health Organisation, Geneva, Switzerland.

Zerobin, W. (2005): *personal communication*

AN INTEGRATED WATER QUALITY SECURITY SYSTEM FOR EMERGENCY RESPONSE

WILLIAM B. SAMUELS* AND
RAKESH BAHADUR
Science Applications International Corporation
1410 Spring Hill Road
McLean, VA 22102 USA

Abstract. A set of waterborne transport tools has been developed that simulate the fate and transport of chemical, biological and radiological contaminants in source water and within the distribution system. These tools consist of: (1), RiverSpill – real-time, time-of-travel and dispersion model, (2) PipelineNet – water distribution hydraulic and water quality model and (3) the Incident Command Tool for Drinking Water Protection (ICWater). RiverSpill is a geographic information system (GIS)-based tool that provides the ability to model, using real-time stream flow data, the time-of-travel and concentration of toxic substances at public water supply intakes. PipelineNet is a GIS-based system, which integrates hydraulic and water quality models with existing spatial databases. PipelineNet integrates the EPANET hydraulic model and ArcView to give emergency managers real time information for estimating the risks to public water supplies. The integrated system calculates, locates, and maps the population at risk from the introduction of contaminants to the water distribution network. ICWater integrates multiple sources of information to give decision makers concise summaries of current conditions and forecasts of future consequences of terrorist acts on public water supply safety. The system is GIS-based and the output is compatible with the Defense Threat Reduction Agency's (DTRA) Consequences Assessment Tool Set (CATS) and the Environmental Protection Agency's Emergency Response Analyzer. The core element of ICWater is the RiverSpill time-of-travel model. RiverSpill is being modified to operate at the 1:100,000-scale stream network available through the EPA and USGS National Hydrography Dataset (NHD).

**To whom correspondence should be addressed. Dr. William B. Samuels, Science Applications International Corporation, Hazard Assessment and Simulation Division, 1410 Spring Hill Road, McLean, VA 22102 USA.*

J. Pollert and B. Dedus (eds.), Security of Water Supply Systems: From Source to Tap, 99–112.

Keywords: water quality, modeling, hydraulics, geographic information systems, consequence assessment, emergency response

1. Introduction

As pointed out in "Protection of Public Water Systems," prepared for the President's Commission on Critical Infrastructure Protection, the United States has approximately 180,000 water systems, serving over 250 million persons (Rycus and Snyder, 2001). An estimated 16 trillion gallons of water is produced annually in the United States by water utilities. This water is supplied by approximately 880,000-mile distribution system, to residential, commercial, and industrial properties. The primary mission of each utility is to efficiently provide their customers with a safe and reliable supply of water. The availability of adequate quantities of water, on demand, at sufficient pressure, and safe for use, has become a public trust. Concern that water infrastructure in the United States can be vulnerable to terrorist attack –biological, chemical, structural, cyber – has been heightened by recent events. Water systems are complex and any component of the system can be compromised with contamination.

Water systems and their components are particularly vulnerable, with each component susceptible to a range of disruptive or contaminate actions. The point of entry of any contaminant to a water distribution system determines its impact on the system. Contamination can be introduced in raw water sources, treatment facilities, or in distribution network. Different models and methodologies analyze fate and transport of the contaminants at different spatial locations. There is no model/system that analyzes fate and transport of contaminants through sources, intakes, treatment, storage, and/or distribution.

This paper introduces the concept of integrated water quality security. Assuring safe and adequate quantities of water, delivered at sufficient pressure is best achieved by an integrated water quality security approach and involves virtually every component of a water system (raw water sources, intakes, treatment plants, pipeline network). The Integrated Water Quality Security System (IWQSS) (see figure 1) consists of the following four components:

- RiverSpill – tracks fate and transport (in real-time) of contaminants in surface water
- Water Treatment Process (WTP) – simulates the water treatment process component by removing any contaminants as per the treatment processes. Percent removal per treatment process can be input or they can be selected from a drop down menu.

- PipelineNet – tracks contaminants within the water distribution system (pipeline network). It can track both, end of pipe contaminants from the water treatment plant and contaminants introduced after water treatment.
- ICWater – Incident Command Tool for Protecting Drinking Water. ICWater integrates the RiverSpill model with a high resolution stream network (National Hydrography Dataset) in a desktop and web-based environment.

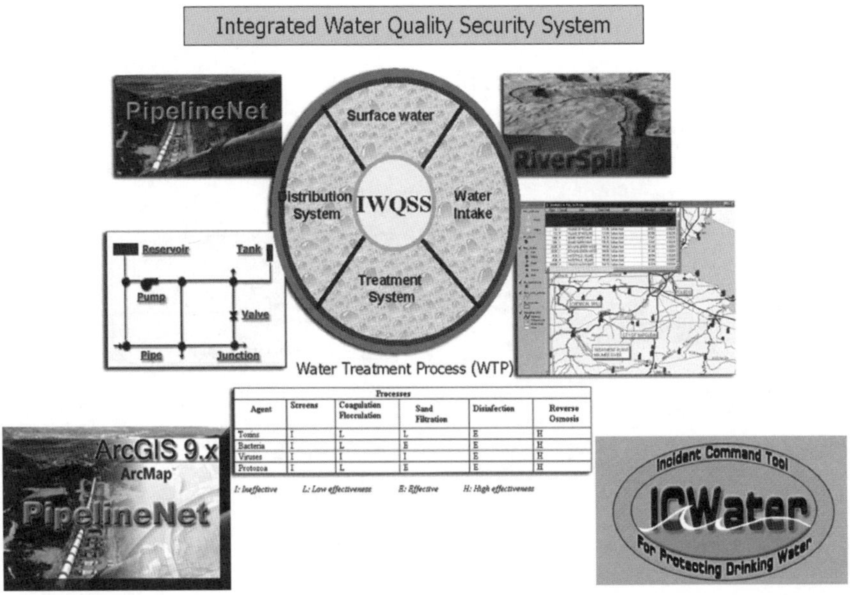

	Processes				
Agent	Screens	Coagulation Flocculation	Sand Filtration	Disinfection	Reverse Osmosis
Toxins	I	L	L	E	H
Bacteria	I	L	E	E	H
Viruses	I	I	I	E	H
Protozoa	I	L	E	E	H

Figure1. Schematic diagram of the Integrated Water Quality Security System (IWQSS)

All the components operate from the same platform, ArcView, a GIS developed by the Environmental Systems Research Institute (ESRI).The IWQSS system was developed for application during the 2002 Winter Olympics in Salt Lake City, Utah. RiverSpill was tested and is operational for the state of Utah and Ohio. It has been expanded to provide coverage of the continental US. The PipelineNet system was tested using the Salt Lake City database composed of approximately 31,000 links and 52 pressure zones. It has also been applied to additional cities in the US. The WTP is embedded in the system and can work as the last step in RiverSpill or the first step in PipelineNet. ICWater is a major upgrade to the RiverSpill model that includes conversion to ArcGIS (version 9), a 1:100,000 scale stream network for the US and compatibility with other emergency response systems.

2. RiverSpill

RiverSpill (Samuels et al., 2002) is a GIS-based software tool with integrated data base capability that is used to track and model the flow and concentration of contaminants in source waters (surface). Figure 2 represents the system architecture for RiverSpill. The RiverSpill system calculates, locates, and maps the population at risk from the introduction of contaminants to the public water supply. The system contains the following models and databases:

- Stream flow and transport models
- USEPA Reach File database
- USEPA Public Water Supplies database
- USGS Real-Time Stream Flow Gage database

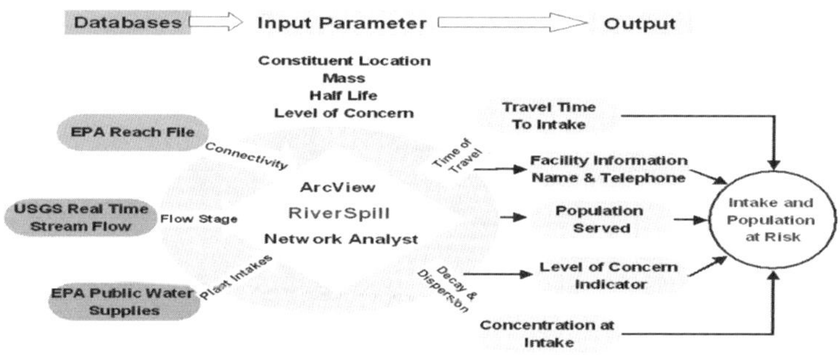

Figure 2. RiverSpill architecture diagram

The databases required for this tool include the Enhanced River Reach File (ERF1-2), the USGS real-time stream flow measurements, and the EPA Safe Drinking Water Information System (SDWIS). The ArcView Network Analyst extension is used to integrate the databases and to provide the user with a tool to quickly assess the consequences of the introduction of a chemical or biological contaminant to the source waters (surface water) of a public water supply.

The original EPA Reach File Version 1.0 (RF1) is a vector database of approximately 700,000 miles of streams and open waters in the conterminous United States. The RF1 stream flow data consist of mean annual flow and low flow estimates made at the downstream ends of more than 60,000 transport reaches. The digital data set ERF1-2 includes enhancements to RF1 to ensure the hydrologic integrity of the digital reach traces and to quantify the time of travel of river reaches and reservoirs. ERF1-2 was designed to be a digital database of river reaches capable of supporting regional and national water-quality and river-flow modeling and transport investigations in the water-resources community.

The stream-gauging program of the USGS is an aggregation of networks and individual stream flow stations that originally were established for various purposes. Approximately 5,000 of the 6,900 U.S. Geological Survey sampling stations are equipped with telemetry to transmit data on stream flow and temperature back to a database for real-time viewing via Internet.

The locations of public water supply plants and intakes were obtained from SDWIS. It contains information about public water systems and their violations of EPA's drinking water regulations. SDWIS is an EPA national database storing routine information about the nation's drinking water. The SDWIS stores the information EPA needs to monitor approximately 175,000 public water systems.

The RiverSpill Model uses a convective-diffusion equation to transport the contaminant with the water; and allow for longitudinal dispersion of the contaminant. Time of travel is based on real-time stream flow measurements. The contaminant is also allowed to decay with time.

Figure 3 shows the RiverSpill output for a hypothetical spill on the Maumee River in western Ohio. Using the Network Analyst, "Find Closest Facility" function, the nearest gage to the spill site is determined. The URL for this gage is extracted from the gage attribute table and passed to an Internet Browser. The real-time flow and stage are displayed in the Browser window. In addition, the flow and stage data are extracted by a parsing program and passed to a flow-velocity algorithm. The time-of-travel for each reach, based on the real-time velocity is calculated. The "Find Closest Facility" is invoked again, using the public water supply intakes as the facilities and the time-of-travel attribute as the cost field. Public water supplies along the flow path are displayed on the map and in an associated table. Population served and personnel responsible for operation of the water supply facilities, and their telephone numbers, are also identified in the table. If the concentration of the pollutant at the intake is above a specified level of concern (i.e. a maximum concentration level, MCL), then the intake and its associated table record are colored red to indicate a warning.

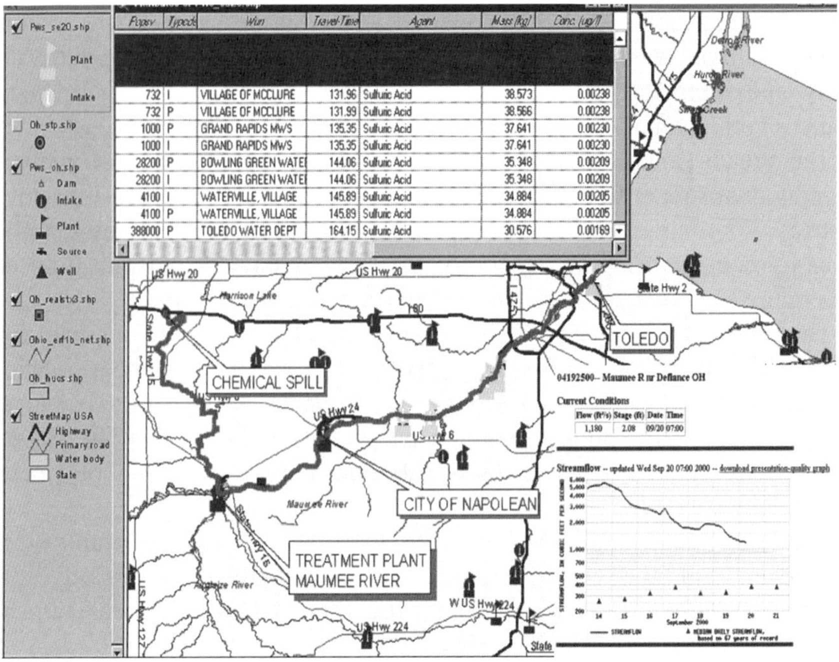

Figure 3. Example output from RiverSpill

3. Water treatment process (WTP) module

There are many chemical and biological agents, which can do harm to a population when introduced into a water supply system (Deininger 2000). Some agents can be credible threats and others are not. For a contaminant to be effective it must meet certain criteria (Deininger and Meier, 2000).

- High toxicity – deadly effect in small amounts.
- Tasteless, odorless, and colorless.
- Chemical and physical stability in water.
- Delayed action to protect the sabotage agent.
- Difficulties in detection with normal analytical methods.
- Unusual effect of poisons; no known antidotes.

Compared to aerial attack (inhalation or skin contact), effective doses are easier to obtain in water (less dilution than air and directly ingested by the target population), and in many cases the materials are more stable (protected from ultraviolet and temperature extremes, although exposed to chlorine) (Hickman, 1999). The effectiveness of a contaminant is maximum only when released after treatment as they would be less susceptible to dilution and would reside in the system for shorter times, thus diminishing the effects of disinfectants and chemical decomposition and oxidation. Various treatment options can be effective for different contaminants.

The major protection against a terrorist attack is the treatment plant processes. They vary in their effectiveness, but a complete treatment plant significantly reduces the contaminant as shown in table 1 (Deininger 2000), and table 2 (AWWA, 1990). Each individual case needs review since no two surface waters are equal, and the treatment processes are not equally efficient, even if they are of the same type. These tables should be used with extreme caution. The data need to be verified with some site or process specific numbers. Systems that have additional treatment options for specialized use (carbon adsorption, ion exchange and membrane filters) may be more successful in reducing the threat but the extent to which that can be quantified is not known at this time.

Table 1. Summary of the effectiveness of surface water treatment processes on major contaminant groups (Deininger, 2000)

Agent	**Screens**	**Coagulation / Flocculation**	**Processes**	**Sand Filtration**	**Disinfection**
Toxins	I	L	**Reverse Osmosis**	L	E
Bacteria	I	L	H	E	E
Viruses	I	I	H	I	E
Protozoa	I	L	H	E	E
			H		

I: Ineffective

L: Low effectiveness

E: Effective

H: High effectiveness

There is no industry wide database that shows the treatment method and its efficiency for each utility in the United States. The IWQSS graphical user interface (GUI) has pull down tables that help in the initial assessment and a quick determination of effectiveness of water treatment. If there are site-specific water treatment numbers (percent removal) available, they can replace the default values of the IWQSS. WTP shows the concentration of the contaminant after it has gone through water treatment processes. The output from the water treatment process is input to the PipelineNet component of the IWQSS.

Table 2. Qualitative contaminant removal process effectiveness for major contaminant groups (AWWA, 1990)

	Bacteria	Viruses	Protozoa	VOC	SOC	TOC	Taste/ Odor
Aeration , Air Stripping	P	P	P	G-E	P-F	F	F-E
Coagulation Sediment / Filtration	G-E	G-E	G-E	G-E	P	P-G	P-G
Lime Softening	G-E	G-E	G-E	P-F	P-F	G	P-F
Ion Exchange	P	P	P	P	P	G-E	-
Reverse Osmosis	E	E	E	F-E	F-E	G	-
Ultra Filtration	E	E	E	F-E	F-E	G	-
Disinfection	E	E	E	P-G	P-G	G-E	P-E
Granular Activ.Carbon	F	F	F	F-E	F-E	F	G-E
Powdered Activ. Carb.	P	P	P	P-G	P-E	F-G	G-E
UV Irradiation	E	E	E	G	G	G	G

P – Poor (0-20% removal)

F – Fair (20-60%)

G-Good (60-90%)

E – Excellent (90-100%)

NA – Insufficient Data

4. PipelineNet

PipelineNet (Bahadur et al, 2003) is a GIS-based software tool with integrated data base capability that can be used to model the flow and concentration of contaminants in a city's drinking water pipeline infrastructure. It contains a pipe network hydraulic model - EPANET (Rossman, 2000), maps, and a US Census population database. Additional infrastructure layers such as hospitals and schools can be overlaid using the GIS component of PipelineNet. The PipelineNet model estimates the population at risk due to the introduction of contaminants in the public water supply and graphically maps this population. Figure 4 shows the PipelineNet architecture for the integration of ArcView and EPANET. The integration is accomplished using the EPANET software toolkit, and GUIs written in Visual Basic and the ArcView programming language (Avenue).

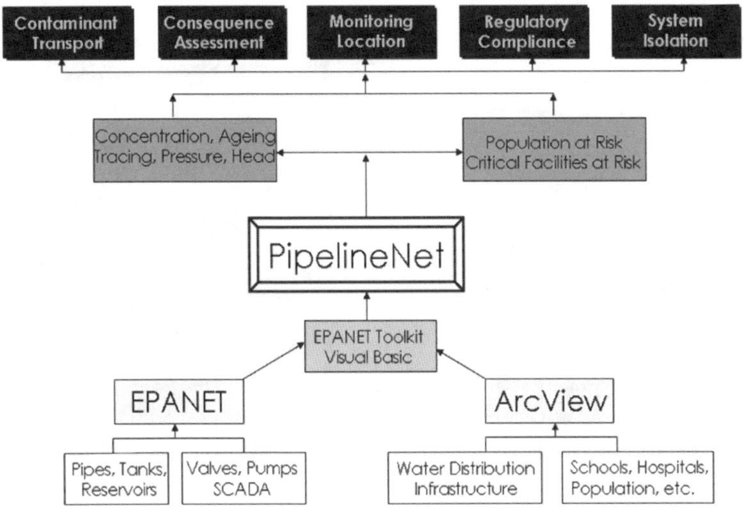

Figure 4. PipelineNet architecture diagram

The first step in applying the PipelineNet system is to prepare an EPANET compatible input file. PipelineNet follows all of EPANET conventions. The EPANET input data file is used to create the PipelineNet spatial databases (locations of pipes, valves, pumps, tanks, etc.) These databases are used for the ArcView display of the output. Creation of these spatial components from the EPANET input file assures synchronization between the GIS and hydraulic databases in PipelineNet.

To initiate a PipelineNet run, a single click on the map anchors an injection point and initiates the EPANET Source Point Parameters GUI. Injection units follow EPANET rules, negative demand nodes can have mass/volume and positive demand nodes can have mass/minute. When PipelineNet is finished processing, the resulting hourly simulation concentrations will automatically load in the ArcView map display and can be viewed as an animation. Pipes change color through time. The color changes reflect varying concentrations of the contaminant.

PipelineNet has all the functionalities of EPANET for simulation (e.g. ageing, tracing, and multiple sources). The PipelineNet output consists of GIS layers that can easily be displayed with other city infrastructure layers (schools, hospitals, roads, etc.). PipelineNet can also output population at risk associated with a contamination level (figure 5). Based on the US Census data a further breakdown in the demographic data (age distribution) is possible.

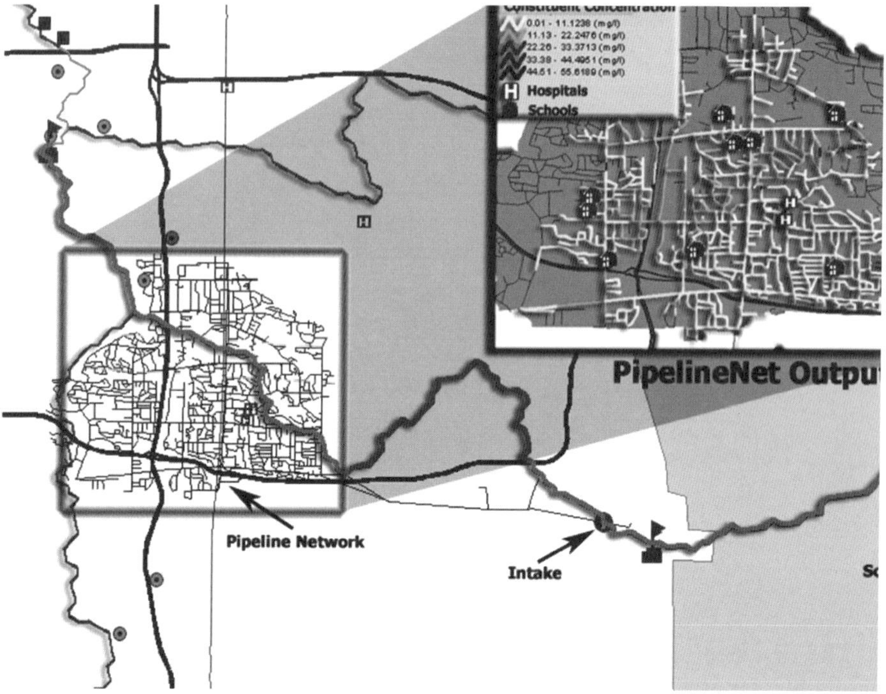

Figure 5. Example PipelineNet output

5. Incident command tool for protecting drinking water (ICWater)

This project involves the development an information tool that will give Incident Commanders the critical information that they need to make informed decisions regarding the consequences of threats to public water supplies. The tool is designed to meet the specific needs of Incident Commanders with respect to its content, timing and spatial coverage and resolution. ICWater, (Samuels et al., 2004) being a multi-disciplinary application, meets the following requirements:

- Integrate all critical data needed to evaluate and respond to an incident into a GIS-referenced system.

- Predict dispersion of waterborne contaminants by integrating the EPA and TSWG RiverSpill tool with the National Hydrography Dataset (NHD)
- Incorporate interfaces between field sensors and the RiverSpill tool
- Develop an interface for input of field reports by first responders and mobile units.
- Incorporate an interface for inclusion of hospital admissions data
- Contain GIS layers and databases to display water threats in relation to: surface water contamination sensor locations; sensor outputs; the location of dams, reservoirs, and locks; the location of surface water bodies; all public drinking water intakes; roads and other terrestrial transportation networks; topography; and population
- Provide secure web-based access to local incident commanders, and to a centralized, regional or national command center.
- Provide the capability of tracking human pathogens, toxic chemicals, and radioactive substances that pose significant threat to public safety in case they were used to attack water sources.

The technical approach to develop an operational incident command system is: (1) integrate existing components to interact seamlessly, (2) upgrade the supporting data bases to give the tool national coverage at the appropriate level of detail, (3) enable the tool to run on the web, and (4) maintain the tool ready to be available quickly in an emergency and for training. As shown in figure 6, the ICWater integrates real-time flow data from USGS gages as well as contaminant detections from field sensors and reports with geographic information (NHD, GIS layers) and the RiverSpill model to provide up-to-date maps, reports and tables that enable emergency response personnel to make timely decisions about people at risk and what actions will most effectively reduce that risk. ICWater is based on three-tier system architecture. Each tier is described below:

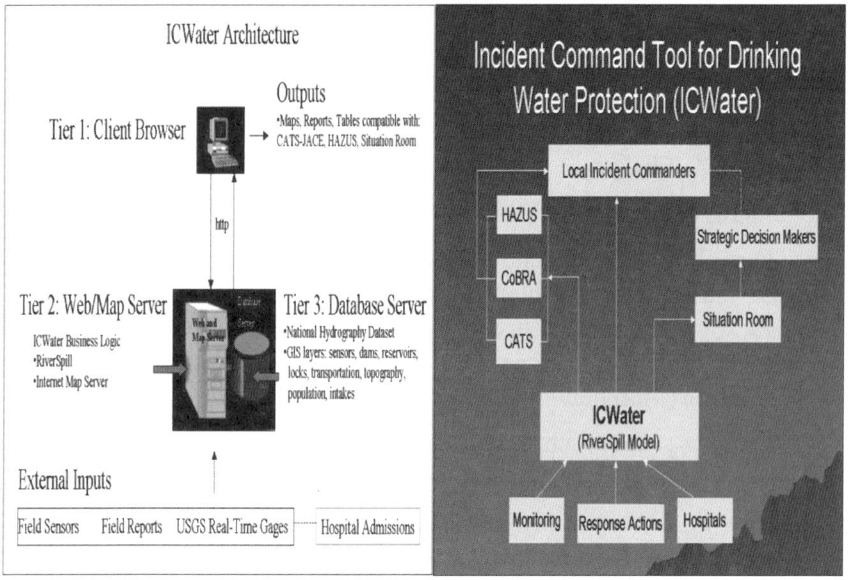

Figure 6. ICWater schematic diagram

- Tier 1: End user machine with an Internet browser, such as Netscape or Internet Explorer. The graphical user interface will allow the user to execute all ICWater functions as well as export maps, tables and reports to CATS, HAZUS, the EPA Emergency Response Analyzer and CoBRA.
- Tier 2: Web and Map Server, which houses the ICWater business logic including the RiverSpill model and the Internet Map Server
- Tier 3: Database Server, which houses the National Hydrography Dataset, and GIS layers

The system accesses external data from the USGS real-time stream gauging stations and contaminant detections from field sensors and reports. Figure 7 shows example output from ICWater for both downstream and upstream tracing.

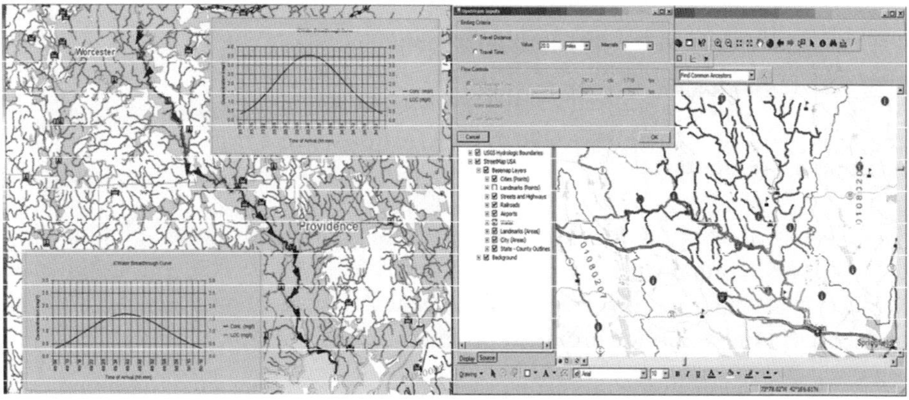

Figure 7. Example ICWater output for downstream (left) and upstream (right) tracing

6. Conclusions

The IWQSS allows the user to:

- Select a location on a river to introduce a chemical or biological contaminant
- Track the contaminant downstream, under real-time flow conditions, to a water supply intake.
- Further track the contaminant to a water filtration plant.
- Simulate the effectiveness of water treatment processes on the contaminant
- Model the transport of the contaminant through a water distribution system.
- Identify the overall population served by the plant and infrastructure impacted.

The advantage of the integrated system is that it can comprehensively study all the components of the system. The IWQSS can track contaminants introduced anywhere in the system:

- Upstream of the intake of a water supply system;

- At the water intake or the treatment plant;
- At a service reservoir;
- At a point in the distribution system;
- Individual house connection

Acknowledgements

The Technical Support Working Group (TSWG) funded the development of the IWQSS components under a contract to SAIC. The Environmental Protection Agency (EPA), Federal Emergency Management Agency (FEMA) and the US Forest Service provided program management.

References

AWWA, 2000. Committee Report: Disinfection at medium-size and large systems, J. AWWA, Vol. 92, No 5, May 2000.

Bahadur, R., Samuels, W.B, and Pickus, J. 2003. Case study for a distribution system emergency response tool, American Water Works Association Research Foundation Report No. 2922, Denver, CO, 50p.

Deininger, R. A., 2000. The Threat of Chemical and Biological Agents to Public Water Supply Systems,Water Pipeline Database, TSWG Task T-1211, SAIC Hazard and Assessment Division, McLean, Virginia, Contract N 39998-00-C-0633, June 2000.

Deininger, R. A., and. Meier, P. G., 2000. Sabotage of public water supply systems, in Security of Public Water Supplies, R. A. Deininger (ed.), NATO Science Series, volume 66, Kluwer Academic Publishers.

Hickman, D. C., 1999. A Chemical and Biological Warfare Threat: USAF Water Systems at Risk, Counter Proliferation Paper No. 3, Future Warfare Series N0. 3, Air War College, Maxwell Air Force Base, Alabama, September, 1999.

Rossman, L.A., 2000. EPANET 2 User's Manual, National Risk Management Research Laboratory, Office of research and Development, USEPA, EPA Publication EPA/600/R-00/057, 200 p.

Rycus, M. J. and Snyder, J.C., 2001. Protection of Public Water Systems: Emergency Response and Water System Security in Germany, Israel, and Japan and Vulnerability of Public Water System Supervisory Control and Data Acquisition (SCADA) Systems, prepared for the President's Commission on Critical Infrastructure Protection by, The Studies in Urban Security Group (SUSG).

Samuels, W.B., Bahadur, R., Pickus, J. Amstutz, D. and Ryan, D. 2004, Development of the Incident Command Information Tool (ICIT) for Drinking Water Protection, Proceedings, Geographic Information Systems and Water Resources III, American Water Resources Conference, May 17-19, 2004, Nashville, TN

Samuels, W.B., Bahadur, R., Amstutz, D.E., and Pickus, J. and Grayman, W. 2002, RiverSpill: A GIS-Based Real Time Transport Model for Source Water Protection, Proceedings Watershed 2002, February 24-27, 2002, Fort Lauderdale, FL

MATHEMATICAL MODEL AS A TOOL TO ENSURE HIGH QUALITY OF DRINKING WATER IN A DISTRIBUTION SYSTEM

KATEŘINA SLAVÍČKOVÁ*, ALEXANDER GRÜNWALD, MAREK SLAVÍČEK, BOHUMIL ŠŤASTNÝ, KLÁRA ŠTRAUSOVÁ
Czech Technical University in Prague, Faculty of Civil Engineering, Dept. of Sanitary and Ecological Engineering, Thákurova 7, 166 29 Praha 6

Abstract. Corrosion of iron pipes negatively influences the quality of distributed water in many aspects. Colour, turbidity, conductivity and iron concentration increase, whereas dissolved oxygen and active chlorine decrease. Corrosion can also cause a failure of pipes and secondary contamination of drinking water. This paper deals with changes of water quality in a distribution network and with the use of the Epanet 2 program for modelling of increasing iron concentrations during the distribution of drinking water. This research is being conducted in cooperation with the operator of the South Bohemian Waterworks providing transport of drinking water from the Plav WTP (Water Treatment Plant) to the Hodušín water tank. The data required for this analysis were measured and evaluated, and the model was set up and calibrated using the average values of the first set of measured data. Different variants of kinetics and coefficients were evaluated and the best of them were verified. The aim of this research is to reduce corrosion and ensure better water quality not only with respect to iron concentrations, but also for other parameters, for example, total active and free chlorine, water stability, and so on. Corrosion rates were also measured and different types of corrosion were detected.

Keywords: water distribution system, water quality modelling, corrosion, coupon tests

*Ing. Kateřina Slavíčková, Ph.D., Czech Technical University in Prague, Faculty of Civil Engineering, Dept. of Sanitary and Ecological Engineering, Thákurova 7, 166 29 Praha 6, Czech Republic, e-mail: katerina.slavickova@fsv.cvut.cz

J. Pollert and B. Dedus (eds.), Security of Water Supply Systems: From Source to Tap, 113–124.

1. Introduction

A model of water quality in a distribution system is an important tool, which helps water utilities to deliver to all consumers' safe drinking water that meets the water quality standards. Models of water quality changes in a distribution system that are successfully calibrated and verified can be used for many purposes, Including, for example, prediction of water quality at any point in the system, optimisation of the distribution system operation with respect to water quality and residence time, evaluation of the effect of the disinfection system, an evaluation of the impact of the stage of distribution system on water quality, and also optimisation of water tank operation from this point of view.

This paper deals with water quality changes during the distribution of drinking water and focuses on possibilities of modelling iron concentration increase and free and total active chlorine modelling by the EPANET 2 software, and on corrosion monitoring by coupon tests. The research results presented herein focused on corrosion problems and water quality changes modelling in a part of the South Bohemian drinking water supply system from the Plav WTP to the Hodušín water tank.

1.1. SOUTH BOHEMIAN DRINKING WATER DISTRIBUTION SYSTEM

The South Bohemian drinking water supply system uses mainly surface water sources. The most important of them is the Římov water reservoir that was built on the Malše river during 1971 – 1978. A water supply system was constructed during 1982-1995. Today it belongs to the largest water supply systems in the Czech Republic. The water supply system capacity was designed according to studies of predicted demands. After 1989, the situation changed and demands are now lower due to reduced use and restructuring in industry and agriculture, and a reduced consumption of water in households. Consequently, only one half of the system capacity is now being used. Water stays longer in the distribution system that is made mainly of steel and water quality is decreasing and iron concentrations are increasing. In total, 250,500 inhabitants are supplied entirely with water from this supply system and another 110,000 inhabitants use this water in combination with local sources. Also, there are further 140 customers from industry and agriculture with consumption higher then 5,000 m^3/d each using this water.

The South Bohemian drinking water supply system is divided in to six parts and presented research was done in part V, spanning from the Plav WTP to the Hodušín water tank.

2. Methods and models

2.1. MODELLING OF WATER QUALITY IN EPANET 2

Epanet 2 is a computer program that performs extended period simulation of hydraulic and water quality behaviour within pressurized pipe networks. A network consists of pipes, nodes, pumps, valves, storage tanks and reservoirs. Epanet 2 tracks the flow of water in each pipe, the pressure at each node, the height of water in each tank and the concentration of a chemical throughout the network during a simulation period comprised of multiple time steps.

Water quality modelling can be used for modelling the age of water throughout the network. Zero order growth is used in Epanet to model water age, where, with each unit of time, the "concentration" (i.e., age) increases by one unit. It also enables to model the movement and changes of reactive material as it grows (e.g., a disinfection by-product) or decreases (e.g., chlorine residual) with time. It is possible to model reactions both in the bulk flow and at the pipe wall with the use of on the order kinetics for reactions in the bulk flow and zero or first orders kinetics for reactions at the pipe wall. Epanet's water quality simulator uses a Lagrangian time-based approach to track the stages of discrete parcels of water as they move along pipes and mix at junctions during the constant time steps.

2.2. CORROSION MEASUREMENT

Monitoring and corrosion measurement is a necessary part of monitoring of stages and quality of the system and its changes. It is used for determining the efficiency of anti-corrosion measures during chemical water treatment. It helps to deduce the optimal conditions for water treatment (temperature, pH, alkalinity), which limit or stop corrosion. Measurement results are used to extend the working life of facilities, improve water quality, predict method and time of maintenance, and thereby help reduce operational costs. Some techniques yield results immediately; others inform us about corrosion rates or about the total corrosion.

Monitoring techniques can be divided into three basic groups:

- *Metal decay measurements* – a physical decay of material caused by the corrosion or an erosion,
- *Electrochemical measurements* – corrosion effects of the environment are measured without viewing the actual decay of a material, and
- *Supporting analyses* – total iron, biological film, pH, etc.; those are laboratory experiments which help understand the system.

2.2.1. *Coupon tests*

A corrosion coupon test is a widely used and relatively simple method for corrosion monitoring. It is based on the accurate measurement of the decrease of weight of iron of exposed samples. The time of exposure is usually 30, 60 or 90 days. The exact technique of coupon tests is written in TNV 75 7121 directive "Requirements on water quality distributed by a pipe network." The tests are carried out with testing coupons made from a steel plate of class 11, of a size of 42 x 42 x 1 mm (width x length x thickness). Each coupon is marked with a number, mechanically cleaned, and then cleaned with an organic degreased agent. It is also cleaned of potential corrosion products by a short staining in solution of hydrochloric acid with hexamethylenetetramine. Finally, it is washed with distilled water followed by ethanol, dried by hot air and then placed in desiccators. After that each coupon is weighed with an accuracy of 0.0005 g and placed in special holders capable of holding five coupons. The holder with coupons is inserted into a vertical pipe connected by inlet and outlet to the water pipeline. Flow, temperature and active chlorine concentration are measured and recorded during the test. At the end of the test coupons are stained again in hydrochloric acid solution to remove all the corrosion products. They are washed with distilled water, ethanol, dried by hot air and then placed in desiccators. Each coupon is cooled and weighted.

Corrosion rate v_u ($\mu m.r^{-1}$) is computed as follows:

$$v_u = 365 \frac{(U_{t2} - U_{t1})}{t_2 - t_1} \tag{1}$$

Where t_1 is the shorter time of exposure (days), t_2 is the longer time of exposure (days), U_{t1} is corrosion decay during the shorter exposure (μm) and U_{t2} is corrosion decay during the longer exposure (μm).

Corrosion decay U_t (μm) is computed by the formula:

$$U_t = \frac{1}{7.68}(k - k_0) \tag{2}$$

where U_t is corrosion decay during the time of exposure t, k is the mean value of corrosion decay for five coupons (g/m^2), k_0 is corrosion decay of five coupons during the blank experiment (g/m^2) and 7.68 is steel density (g/m^3).

Corrosion decay of individual coupons k (g/m^2) is calculated as:

$$k = \frac{(m_0 - m_t)}{S} \tag{3}$$

where m_0 is weight of coupon before exposure (g), m_t is weight of coupon after exposure (g), and S is total surface of coupon before exposure (m^2).

3. Results and discussion

3.1. WATER QUALITY CHANGES MODELLING IN EPANET 2

3.1.1. *Data collection*

A hydraulic model has been created at our department for the water distribution system from the Plav water treatment plant to the water tank Včelná, Hlavatce, Sudoměřice, and the Hodušín water tank. It was used for residence time modelling and water quality modelling. The total length of this water distribution network is 89 km and the hydraulic model contains 693 nodes, 7 distribution reservoirs, 661 pipes and 21 pumps. Hydraulic model is composed of gravity water conduits –Včelná – Hlavatce, Zdoba - Malá Varta and Malá Varta – Sudoměřice. The water is pumped in parts Plav – Včelná, Hlavatce - Zdoba and Sudoměřice – Hodušín. Water conduit diameters are 1000 and 500 mm.

The sources of water include the Římov reservoir (withdrawal 1,480 l/s), Malše River (maximum intake 1,000 l /s) and the Vidov bore (intake 40 l/s). This water is treated at the Plav water treatment plant that has a capacity of 1,400 –1,600 l/s. It is designed as a two-stage treatment plant with settling tanks and filters. Treated water from Plav WTP, which enters the distribution system, continually meets the national standards and also the EU directive No.98 for all parameters. But the quality of water changes during distribution and there were problems noted in this distribution system, mainly with iron concentrations. This research project was initiated in order to help remedy such problems.

The research work and all measurements were carried out in co-operation with the South Bohemian Waterworks in The town of České Budějovice. Firstly all measured data of iron concentrations were evaluated. Then more detailed measurements of iron concentrations, and free and total active chlorine concentrations were measured during the week from September 9, 2002 to September 13, 2002. These data were used for calibrating the models of water quality changes. Further all necessary measurements for verification of models were conducted from October 2002 to January 2004. Measurements of water quality parameters were done at all sampling points in late 2002, and in 2003 and 2004 at a reduced number of sampling points.

3.1.2. *Chlorine decay modelling*

Different models of free and total chlorine decay were compiled in Epanet 2 using the kinetics of zero and first orders both in the bulk flow and at the pipe wall. Average values of free and total active chlorine concentrations measured during the week from September 9, 2002 to September 13, 2002 were used for calibration of models. Models with the best agreement between the measured and modelled values were verified.

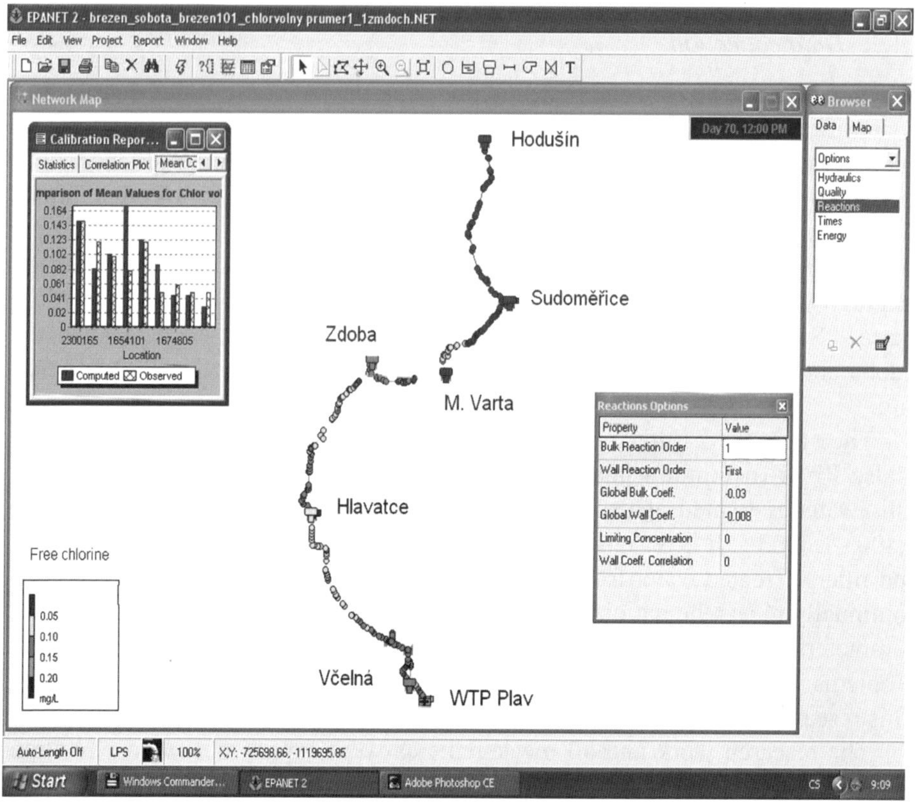

Figure 1. Schematics of the modelled part of the water supply system Plav WTP – Hodušín in Epanet 2, displaying the results of free chlorine modelling

Reaction rate constants are specific and different for each distribution system. These constants are the results of calibration. Different variants of kinetics and coefficients were evaluated, and the following two models of total active chlorine performed the best: model of the 1st order in the bulk flow and 1st order at the pipe wall with the bulk reaction rate constant k_b = -0.1 d^{-1} and the wall reaction rate constant k_w = - 0.06 m/d, and the model of the 1st order in the bulk flow and zero order at the pipe wall with a bulk reaction rate constant k_b = -0.18 d^{-1} and the wall reaction rate constant k_w = - 2.2 $mg/m^2/d$. These two models showed relatively high correlation of measured and computed values and were used for verification.

The results of verifications show that total active chlorine decay can be predicted by these models, but for some other stages of the system and chlorination at Hlavatce and Zdoba tanks, it is more suitable to use the model of the 1st order in the bulk flow and the 1st order at the pipe wall, and for other stages the model of the 1st order in the bulk flow and zero order at the pipe wall, because of chlorine data variability.

Models of free chlorine decay were also calibrated for the values of free chlorine concentrations measured during the calibration week in September 2002. Different variants of kinetics and reaction rate constants were evaluated and the best results for free chlorine were found for the following two models: the model of the 1st order in the bulk flow and 1st order at the pipe wall with the bulk reaction rate constant $k_b = -0.03\ d^{-1}$ and the wall reaction rate constant $k_w = -0.008$ m/d, and the model of the 1st order in the bulk flow and zero order at the pipe wall with the bulk reaction rate constant $k_b = -0.09\ d^{-1}$ and the wall reaction rate constant $k_w = -0.075\ mg/m^2/d$. These two models showed relatively high correlation of measured and computed values and were used for verification.

The results of verification show that free chlorine decay can be predicted by these models and for one third of system stages and chlorination at tanks Hlavatce and Zdoba it was more suitable to model chlorine decay at the pipe wall by 1st order kinetics and for two thirds of days by zero order kinetics because of chlorination in tanks and free chlorine data variability.

3.1.3. *Modelling increasing iron concentrations*

Different models of iron concentration increase were compiled in Epanet 2 using the kinetics of zero and first order both in the bulk flow and at the pipe wall. Average values of iron concentrations measured during the week from September 9, 2002 to September 13, 2002 were used for calibrating models. Models with the best agreement between the measured and modelled values were verified. Mean values can describe water quality changes much better than instantaneous values, because they are not so influenced by sudden changes of conditions and concentrations or by incorrect measurements. Evaluation of measured data of iron concentrations used for calibration of models is presented in Fig. 2.

The best agreement between the measured and modelled iron concentrations during calibrations of models with different kinetics was achieved by the iron concentration increase model of the 1st order in the bulk flow and zero order at the pipe wall with the bulk reaction rate constant $k_b=0.0058\ d^{-1}$ and the wall reaction rate constant $k_w = 1.905\ mg/m^2/d$. An example of the calibration results is shown in Fig. 3.

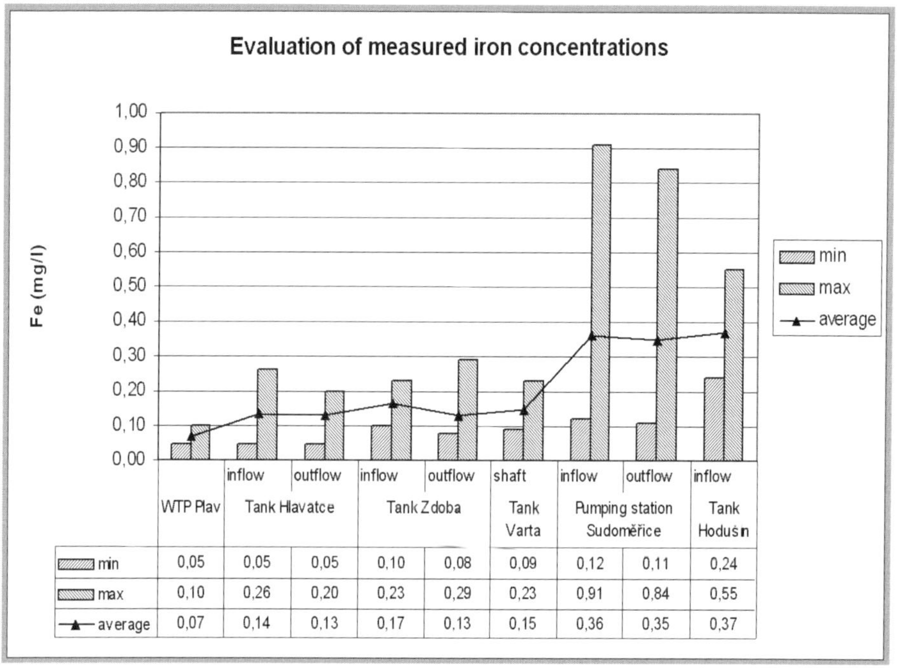

Figure 2. Evaluation of iron concentration measured for model calibration

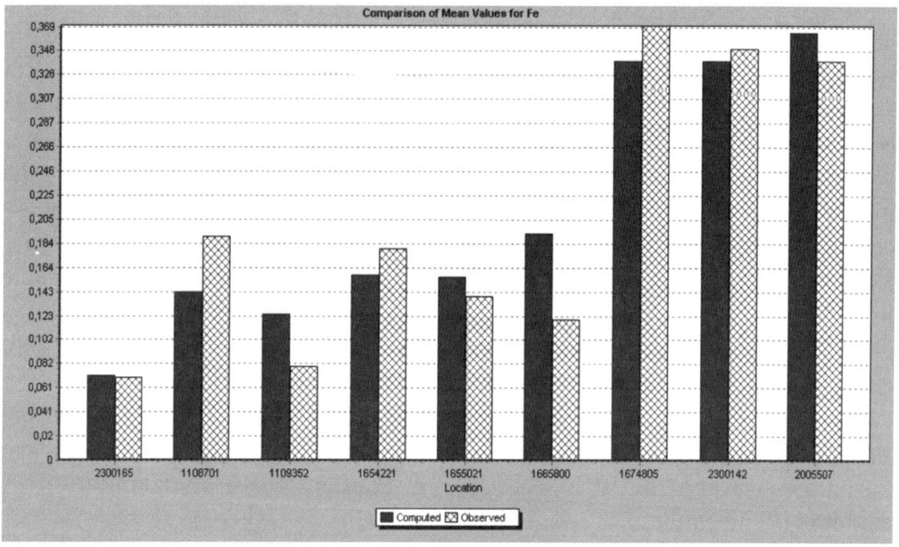

Figure 3. Comparison of means of measured and modelled iron concentration data at important sampling points

On the basis of the calibration results and for research purposes three models of iron increase in the distribution system with different kinetics in the bulk flow and at the pipe wall were chosen for verification. During 2002, all measurements were collected at 9 sampling points in the distribution system Plav WTP – Hodušín tank and from the beginning of 2004; the number of points was reduced to 4. This research focusing on corrosion problems was carried out for five years and some results were used to improve the situation. Measured data from 2003 show that iron concentration values at the end of the system and at the problematic pumping station Sudoměřice have decreased to one half. Verification of models proved that Epanet 2 is a suitable tool for iron concentration increase modelling and that verified models can be used for prediction and system analysis. The residence time and water quality parameters change during the day and also over a longer period depending on the regime of pumping stations.

3.2. RESULTS OF COUPON TESTS

Coupon tests were done with steel coupons in raw, filtered and treated water from the Plav WTP. Evaluation of results is in Table 1.

Table 1. Coupon tests evaluation

	Raw water		Filtered water		Treated water	
Time of exposure (d)	35	70	35	70	35	70
Corrosion decay (k) ($g.m^{-2}$)	52.8	85.3	122.5	167.9	45.4	47.7
Corrosion decay (U_t) (μm)	6.7	10.8	15.6	21.4	5.8	6.1
Corrosion rate (v_u) ($\mu m.r^{-1}$)	70.0	56.6	162.5	111.4	60.2	31.6
Corrosion rate (35 – 70 d) ($\mu m.r^{-1}$)	-	42.8	-	60.5	-	3.1

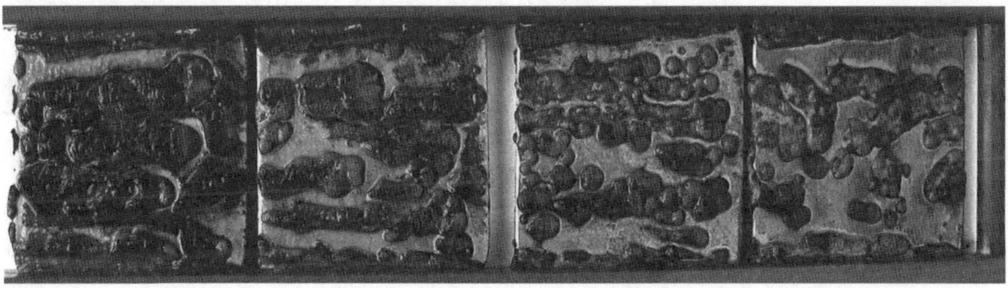

Figure 4. Filtered water – coupons after removal from water – March 8, 2005 (exposure 70 days)

We have measured the corrosive characteristics of treated water and water at chosen sampling points in the distribution system from the Plav WTP to the Hodušín tank. The aim of this research was to decrease corrosion and obtain better water quality not only with respect to iron concentrations but also for other parameters, for example total active chlorine and free chlorine, water stability and so on. Measurements of corrosion rates were also done and different types of corrosion were detected by coupon tests. An example of both dotted and area corrosion is in Fig. 6.

Corrosion of iron pipes negatively influences the quality of distributed water, which is enriched by iron, an inner side of pipe is corroded and incrustations are formed and the content of dissolved oxygen and active chlorine decreases. Corrosion can be also a cause of pipe failure and secondary contamination to drinking water.

Figure 5. Corrosion loop - coupons adjustment

Figure 6. Corrosion coupon after finishing the test – an example of dotted and area corrosion

4. Conclusions

Free chlorine decay and total active chlorine decay were modelled by Epanet 2. In both cases the two best-suited models (the model of the 1st order in the bulk flow and 1st order at the pipe wall and the model of the 1st order in the bulk flow and zero order at the pipe wall) were successfully verified. It is possible to use them for analysing the system, chlorine concentration prediction and for disinfection optimisation in the distribution system.

We concentrated our efforts mainly on corrosion, its monitoring and on ways how to limit corrosion. One part of this work was to confirm that not only chlorine decay, but also iron increase in the distribution system can be modelled in Epanet 2. It was proved by this research.

Models of different kinetics of iron increase were evaluated. The best agreement between the measured and modelled iron concentrations during the calibration process and verification with data from 2002 was achieved by the model of the 1st order in the bulk flow and zero order at the pipe wall with the bulk reaction rate constant k_b=0.0058 d^{-1} and wall reaction rate constant k_w = 1.905 $mg/m^2/d$. Iron concentrations decreased significantly in 2003. Better results for changed conditions were achieved by using model of the 1st order in the bulk flow and 1st order at the pipe wall with bulk reaction rate constant k_b= 0,0035 d^{-1} and wall reaction rate constant k_w=0,0180 m/d. These values were verified by an independent data set.

Hydraulic characteristics, flow and pumping in reservoirs are different for each distribution network and so are the reaction rate constants. Hence model parameters shown here are suitable only for the distribution system WTP Plav-Hodušín. To make simulations more accurate it is necessary to continue this research and find additional factors which influence iron concentration growth during water distribution.

Results of the first set of corrosion tests show that the corrosion rate in water after filtration is twice that of raw water with exposures of 35 and 70 days, respectively. Corrosion rate values were lower in stabilised treated water than in raw water. Residual content of iron in filtered and treated water was under 0,05 $mg.l^{-1}$ due to high pH values.

Acknowledgement

This research was done in cooperation with JčVaKČeské Budějovice. It has been supported by NAZV grant 1G46036 and by MSMT research program MSM 6840770002.

References

Grünwald, A. - Fošumpaur, P. - Šťastný, B. - Slavíčková, K. - Slavíček, M. - Štrausová, K. - Čiháková, I.: Inovace procesu úpravy vody a zabezpečení vysoké kvality pitné vody v distribučních systémech [Výzkumná zpráva]. Praha: ČVUT, Fakulta stavební, katedra zdravotního a ekologického inženýrství, 2004. 1G46036. 62 p. (in Czech).

Grünwald, A. - Mach, M. - Slavíčková, K. - Slavíček, M. - Šťastný, B. - Kasal, R. - Štrausová, K.: Výzkum efektu úpravy vody na její jakost při prodlužujícím se zdržení v rozvodné síti [Výroční zpráva]. Praha: ČVUT, Fakulta stavební, katedra zdravotního a ekologického inženýrství, 2004. QD1003. 80 p.(in Czech).

Grünwald, A. – Mach, M. - Šťastný, B. - Slavíček, M. - Slavíčková, K., Šaršeová, M. – Kasal, R. – Štrausová, K.: Výzkum efektu úpravy vody na její jakost při prodlužujícím se zdržení v rozvodné síti [Výzkumná zpráva]. Praha: ČVUT, Fakulta stavební, katedra zdravotního a ekologického inženýrství, 2003. QD1003. 44 p.(in Czech).

Hubáčková, J. a kolektiv: Výzkum efektu úpravy vody na její jakost při prodlužujícím se zdržení v rozvodné síti. Výroční zpráva 2002, projekt QD 1003, Grantová agentura Mze ČR, VÚV T.G.M., Praha.(in Czech).

Rossman, L.A.: EPANET – User's Guide. 1994.

TNV 75 7121 Požadavky na jakost vody dopravované potrubím, Hydroprojekt cz, a.s., Praha, 2002 (in Czech).

EVALUATION AND MONITORING OF SNOW COVER WATER RESOURCES IN CARPATHIAN BASINS USING GEOGRAPHIC INFORMATION

GHEORGHE STANCALIE*, SIMONA CATANA, CRISTIAN FLUERARU
National Meteorological Administration (NMA), Bucharest, ROMANIA 031686

Abstract. The surveillance and management of water resources in the Carpathian basin of Romania involve the knowledge of snow pack evolution and the snow cover water equivalent estimation in different phases of the snowmelt-runoff season. The paper describes the practical methods developed in the Remote Sensing and GIS Laboratory of the National Meteorological Administration and presents the results obtained for basin snow cover area extent and snowline elevation determination, new snowfall identification, snow cover depletion curves, etc. The method for the evaluation of the water resources stored in the snow pack is based on the use of data from two investigation levels: satellite images and ground information. A complex geo-referenced database, consisting of medium and high resolution satellite data (NOAA-AVHRR, TERRA MODIS, LANDSAT-TM, SPOT-XS), spatial terrain information derived from Digital Elevation Models (DEM) and other exogenous data (maps and ground measurements) was designed and implemented on PC-based computing systems. The GIS developed for the snow cover water resources management includes information on topography, land vegetation cover, land use, soil type, and hydro-meteorological parameters that could be used as separate layers or be interconnected in order to extract the necessary information for a correct and accurate estimation of the snow pack conditions and snow cover water storage during the winter-spring period. This methodology was applied to the main Carpathian catchments in Romania and two examples for the Arges and Bistrita catchments are presented The results obtained using this combined method allowed the permanent monitoring of the

* To whom correspondence should be addressed. Gheorghe Stancalie, National Meteorological Administration, Remote Sensing & GIS Laboratory, 97, Soseaua Bucuresti-Ploiesti, Sector 1, 013686 Bucharest, Romania; e-mail: gheorghe.stancalie@meteo.inmh.ro

J. Pollert and B. Dedus (eds.), Security of Water Supply Systems: From Source to Tap, 125–135.

snow pack dynamics during the winter-spring seasons and to compute the water volume and the mean depth of water equivalent, with good precision.

1. Introduction

The management of the water resources in Romania represents an issue of great importance considering the fact that the natural potential of the water courses is rather limited, and altered by water consumers and by those who redistribute the flow in time. In this respect, the efficient and rational management of water resources also involves the estimation of water resources stored in the snow pack in the Carpathian basin (*Stancalie et al.,* 2000).

The evaluation of the snow cover water storage has to take into account the influences of geographical factors (location, relief, and land cover/land use), the meteorological parameters (air temperature, precipitation, wind speed and direction), the snow cover characteristics and evolution (snow pack depth, snow cover extent, snow density and water equivalent).

The use of satellite images and spatial terrain information proved to be well suited for establishing relationships between a hydrological state vector and the measurable interdependent features of the catchments (characteristics of the relief, hydrological issues, vegetation types and density) (*Martinec & Rango*, 1986).

2. Data used and study areas

The data used in the assessment and monitoring of the snow pack characteristics are provided at two levels of investigation: satellite/aerial photos and ground information. This approach is necessitated by:

- the complex physiographical peculiarities of the Carpathian catchments and their relatively limited areas which call for the use of large scale imageries;
- the necessity of determining both the extent of the snow cover and its condition which calls for the use of multi-spectral image-data;
- the instability in time, during the winter-spring season, of the maximum stored amount and the start of the snowmelt process which requires a continuous surveillance of the snow cover imposing the need for repetitive air and satellite imageries and data from the hydrometeorological network in rapid flux; and
- the need to correlate remotely sensed data with the ground truth.

The snow cover water storage evaluation and monitoring in the Carpathian basin of Romania are based on geo-coded spatial information organized in a

GIS database: data referring to the topography (organized in a DEM), hydrographical network, land vegetal cover / land use, and soil types.

The DEM is very important for the evaluation of the morphometrical parameters of the basin: the slope and aspect, the exposure of the slopes, the slope limit variation, etc. The intersection of the DEM with the vegetation cover layer allows, for example, the determination of forest areas, distributed on altimetry stages.

The geo-referenced information may be used as separate or interconnected layers in order to extract useful information for a correct and accurate estimation of snow pack conditions and snow water storage during the winter-spring period.

The possibility of merging LANDSAT-TM/ETM+, SPOT, ASTER, etc. satellite multi-spectral imagery with the DEM allowed updating the catchments land cover / land use and of other morph metrical parameters, which influence snow accumulation and snow melting processes.

The study areas for the evaluation of the snow cover storage are situated in the main Romanian hydrographical catchments serving hydro power generation interests associated with retention lakes, such as Arges (Vidraru Lake), Lotru (Vidra Lake), Ialomita (Bolboci), Dambovita (Pecineagu), Somes (Fantanele), Bistrita (Izvorul Muntelui Lake), etc.

The methodology for determining the snow water storage in the hydrographical Carpathian catchments of interest is presented in figure 1.

2.1. MONITORING THE SNOW ACCUMULATION PROCESS USING FAST FLUX DATA FROM A HYDRO METEOROLOGICAL STATION NETWORK

The data concerning the depth and density of the snow cover are obtained from the measurements carried out within the national hydrometeorological network, starting from the month of January of the current year, until the middle of May. A new technology has been developed by the Remote Sensing and GIS Laboratory in co-operation with the Computation Center of the National Meteorological Administration for undertaking and processing data supplied in fast flux.

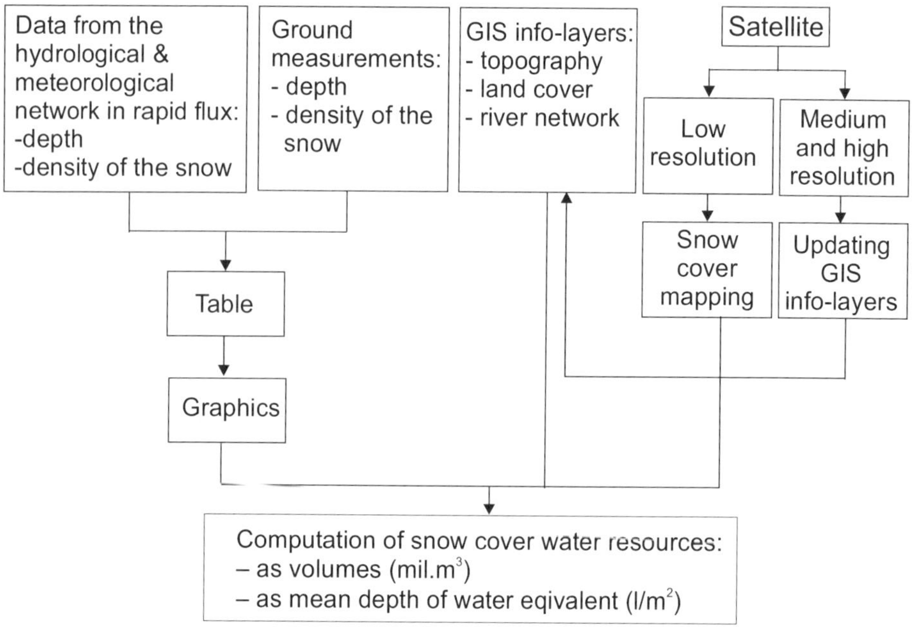

Figure 1. Methodology for determining snow water storage in the hydrographical Carpathian catchments of interest

The general purpose of the application of "ZAPADA" is to operatively exploit daily data transmitted as PLUVIO telegrams. From the records transmitted in these telegrams, two types of data are used: liquid precipitation data (daily maximum and monthly accumulation) and data concerning precipitation in the form of snow recorded over the territory, which are daily transmitted to the Remote Sensing & GIS Laboratory in a dedicated file from the intranet computer network. These data are extracted from the daily meteorological database and then subject to relatively complex processing. This process's structure is imposed by the complexity and specific properties of the data, and by their transmission, undertaking and processing modes.

2.2. PROCESSING AND ANALYSIS OF MEASURED DATA FOR THE DETERMINATION OF THE ALTITUDE LIMIT OF SNOW COVER AND ITS DIFFERENT WATER STORAGE CONDITIONS

Information collected in the field with respect to the characteristics of the snow cover, is correlated with fast flux data coming from meteorological stations (allowing to continually monitor occurrence, storage and depletion of snow cover), the created GIS layers and with the remotely sensed observations of the snow covered area.

The collected data, in a graphical form (Figure 2), represent the dynamics of the snow cover during the winter and spring months at the main meteorological stations within the hydrographic catchments studied or in the areas adjacent to these basins.

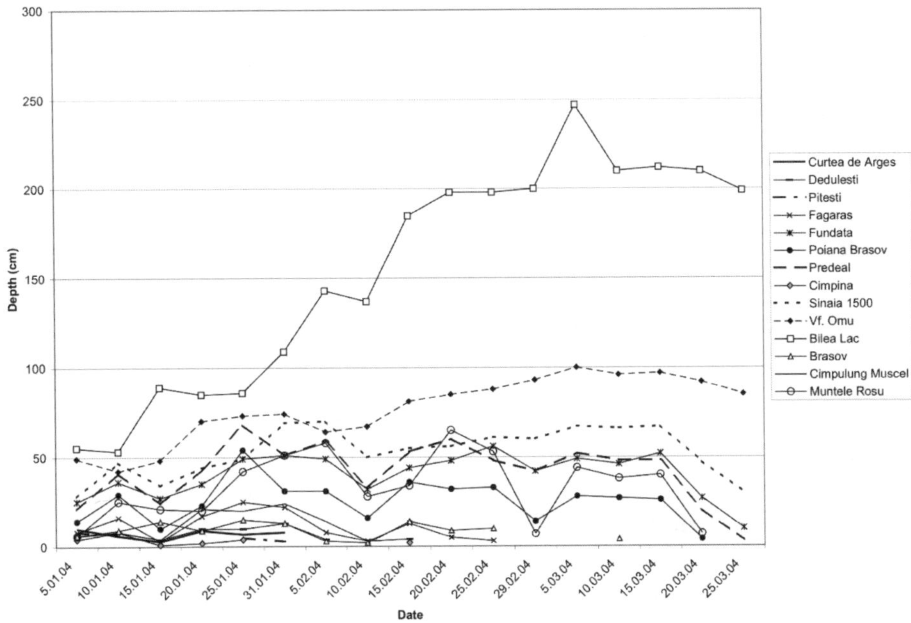

Figure 2. The evolution of the snow depth during the winter-spring season 2003-2004 in the Arges catchments

2.3. DESIGNING THE SAMPLING PROGRAM FOR REPRESENTATIVE PROFILES WITHIN THE CATCHMENTS

Criteria for choosing the ground sampling areas have to take into account:

- the high degree of morphometrical and physiographical features variation (terrain slope and exposure, land cover);
- the land surface geometry and the nature of its coverage; and
- the possibility of clear and fast identification in satellite images.

Also it is necessary to include the morphological profiles, which successively follow the altitudinal stages (both forested and deforested lands), and the separate function of the vertical variation of the meteorological factors conditioning the accretion and ablation of the snow cover.

3. Work method

3.1. CARRY OUT "IN SITU" OBSERVATIONS AND MEASUREMENTS CONCERNING DEPTH, DENSITY AND STATE OF SNOW COVER AND ADJACENT SURFACE

During the period considered to represent the maximum snow storage, a complex program of field observations and measurements starts, including: hydrometeorological parameters,, radioactive characteristics of snow, snow cover characteristics affected by the site topography (altitude, slope, exposure) and by the vegetal cover (coniferous, deciduous or mixed forests, pastures, rocky areas), and water levels in lakes within the studied catchments.

The obtained data lead to production of graphs representing the dynamics of the snow cover during the winter and spring months at the main meteorological stations within the studied hydrographic catchments or the areas adjacent to these basins.

3.2. CREATING A DEDICATED GIS DATABASE

For each study catchments, the geo-coded information is organized in a GIS database:

- data referring to the terrain topography;
- hydrographical network and land cover/land use (for example the info-layers of Bistrita catchments (*Mihailescu,* 2001);
- soil and pedological data;
- municipalities, roads, railways; and
- the hydro-meteorological station network.

3.3. USE OF REMOTELY SENSED DATA TO DETERMINE THE SPATIAL DISTRIBUTION OF THE SNOW COVER

For hydrological simulation and forecasting, estimation of the spatial-temporal variability of watershed parameters, like albedo, soil moisture and snow cover, is very important. The information needed cannot be obtained only from ground survey data, as the number of sampling sites required and the frequency of measurements would be prohibitive. Satellites are well suited for measurements of snow cover because the high albedo of snow presents a good contrast with most other natural surfaces except clouds. Because of this characteristic, snow was observed in the first image obtained from the TIROS-1 (Television and Infrared Observation Satellite) following its April 1960 launch.he satellite data offer valuable information serving to:

- Locate the areas of concern in the Carpathian catchments in remotely sensed data;
- identify areas with snow through differentiation from other bodies in the image having close spectral reflectance, especially the clouds; and,
- delineate and monitor the snow-covered areas and their evolution.

The achievement of the mentioned objectives, involve the processing of satellite images with high temporal resolution, but coarse spatial information (like NOAA-AVHRR or MODIS) using the visible and near IR channels (*Schjødt-Osmo & Engeset*, 1997). The main procedures consist in contrasts enhancement, and binary/multi-threshold segmentation of images.

Spectral reflectivity of snow depends on grain size and shape, impurity content, liquid water content, depth, surface roughness and solar elevation angle (*Hall and Martinec 1985*). Reflectance of fresh snow is very high in the visible wave lengths, but decreases in the near-infrared wave lengths, especially as grain size increases. Fresh snow can have a reflectance up to about 80%, but its reflectance may decrease to below 40% after snow crystals metamorphose.Snow and cloud discrimination techniques are based on differences between cloud and snow/ice reflectance and emittance properties. Clouds typically have high reflectance in visible and near-infrared wave lengths, while reflectance of snow decreases in shortwave infrared wave lengths. The figure 3 presents an example of the snow distribution changes in the Carpathian mountain catchments between March 2003 (A) and June 2003 (B) obtained from the MODIS satellite data with a 250 m resolution.

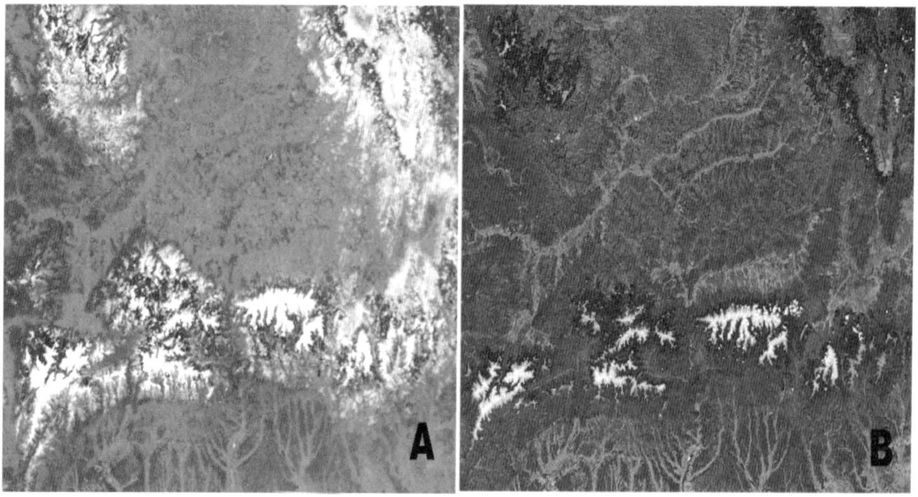

Figure 3. The Carpathian mountain catchments: the snow distributions in March 2003 (A) and June 2003 (B) obtained from the MODIS images

LANDSAT 7 Etm+ data provide a high spatial and spectral resolution imagery set useful for:

- Estimation of snow cover extent expressed as fractions per pixel;
- Estimation of tree canopy cover; and,
- Other geographical and vegetation parameters.

The LANDSAT bands used in this study for snow discrimination are: band 5 (red), band 4 (green) and band 2 (green-yellow). Rocks are reflective, while snow and ice are absorptive in shortwave infrared band 5. Vegetation and snow are reflective in near-infrared band 4. Snow is also reflective in visible band 2, but most rocks and vegetation are absorptive. Using the band-filter combination 5, 4, 2, for color composite, snow appears very clear in cyan, bare rock or soil in magenta, and vegetated areas in green. Mixtures of these land classes cause intermediate colors. Open water is highly absorptive in all bands, and appears black. Clouds are reflective in all bands, and usually appear white, gray or pink.

The results obtained by the un-supervised classification of the LANDSAT 7 ETM+ images allow determination of the snow cover surface. The high-resolution satellite data have been used for the vegetal cover updating, especially of the forest areas, which have an important role in the storage of snow and snow melting processes. The mapping of snow cover becomes limited in areas where snow cover is obscured by forest canopies. A forested landscape is never completely snow-covered because tree branches, trunks and canopies may not be covered with snow. Even in a continuously snow-covered area, much of the forested landscape will not be snow-covered. Furthermore, snow that falls onto the ground through the canopy may not be visible by the airborne or spatial instruments.

The updating of the satellite-based basin land cover has taken into account the advantages offered by the ERDAS Imagine software.

4. Obtained results

This assessment of the snow cover water storage is carried out by altitude stages for various morphometric conditions (slope, aspect) and land categories - forested and deforested, thus yielding: the water volume and the mean depth of water equivalent, for each stage as well as the whole basin (or sub-basins where applicable) for the data considered to represent the maximum snow storage. In Table 1, the results of the snow storage estimates, for each sub-basin of the upstream section of the Arges catchments (Fig. 4), for February 25, 2004, are presented.

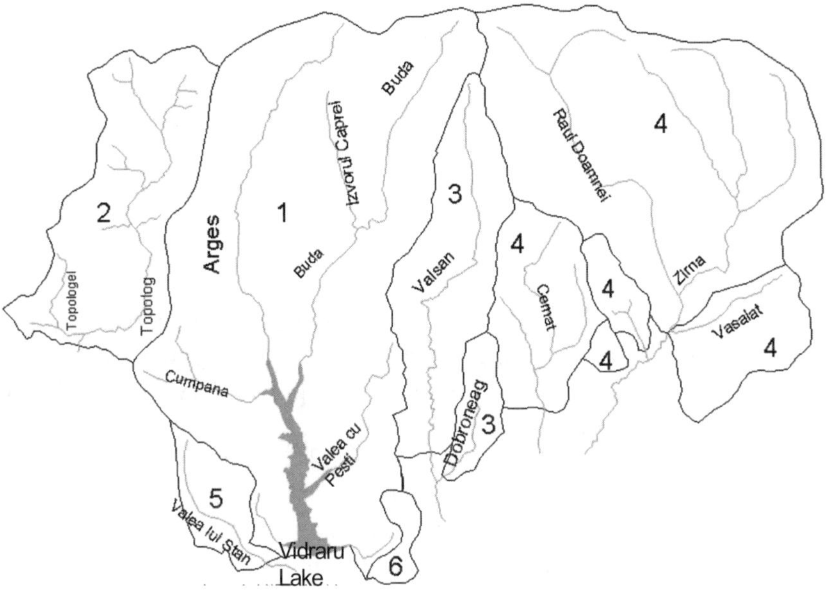

Figure 4. Sub-basins of the upstream section of the Arges catchments

Table 1. Arges catchment – snow cover water storage estimates (February 25, 2004)

				TOTAL	MEAN
	Name of the catchments	TOTAL SUB-BASIN SURFACE	FORESTED SUB-BASIN SURFACE	WATER VOLUME in SUB-BASIN	DEPTH of WATER EQUIVALENT in SUB-BASIN
		(Km^2)	(Km^2)	(mil. m^3)	(m/m^2)
1	Arges main basin (including Vidraru Lake)	283.41	195.28	52.945	0.187
2	Topolog Sub-basin	87.92	53.76	20.309	0.231
3	Valsan+Dobroneag Sub-basin	85.41	52.17	19.322	0.226
4	Raul Doamnei+Bradului+ Draghina+Cernat Sub-basin	256.77	163.49	62.392	0.243
5	Valea lui Stan Sub-basin	19.09	17.45	2.077	0.109
6	Limpedea Sub-basin	7.89	6.87	0.807	0.102
Total		**740.49**	**489.02**	**157.852**	**1.098**

5. Conclusions

In order to extract the useful information for a correct and accurate estimation of the snow pack conditions and of the snow water storage, the most efficient solution is the combined use of remote sensing and GIS techniques. The remotely sensed data offer important advantages such as: uniform data collection and storage, good spatial and temporal resolutions, data availability in digital format, availability of data in less accessible areas, where the acquisition systems do not interfere with the process and the observed phenomena, and the possibility of updating the vegetal cover. The GIS techniques allow taking into account different data referring to topography (organized in DEM), land vegetal cover, land use, soil types, and meteorological parameters, organized in separate or interconnected layers.

The results obtained using this combined method allowed the continuous permanent monitoring of the snow pack dynamics during the winter and spring seasons and to compute, with a good precision, the water volume and the mean water depth stratum in the main Carpathian catchments in Romania.

References

Hall DK, Martinec J. 1985. *Remote Sensing of Ice and Snow*. Chapman and Hall Ltd.: New York; 189.

Martinec, J. & Rango, A. 1986. *Parameter values for snowmelt runoff modeling*, Journal of Hydrology, 84, 197—219.

Maurer, E., Rhoads, J., Dubayah R., Lettenmaier, D. *Evaluation of the snow-covered area data product from MODIS*, HYDROLOGICAL PROCESSES *Hydrol. Process.* 17, 59–71 (2003) Published online in Wiley InterScience (www.interscience.wiley.com). DOI: 10.1002/hyp.1193.

Mihailescu, I.F. 2001. *Studiu climatic si microclimatic al vaii raului Bistrita in sectorul montan, cu lacuri de acumulare*, Editura ExPonto, Constanta.

Schjødt-Osmo, O. & Engeset, R. 1997. *Remote Sensing and snow monitoring: application to flood forecasting*, Norwegian Water Resources and Energy Administration, Hydrological Department, Norway.

Stancalie, G., Catana, S. & Alecu, C. 2000. *Evaluation and monitoring of snowcover water resources in Carpathian basins by remote sensing and GIS techniques*, Proc of the 26-th International Conference on Alpine Meteorology, Innsbruck, Austria.

DEVELOPMENT OF A COMPLEX SYSTEM FOR PIPELINE DESIGN IN SLOVAKIA

ŠTEFAN STANKO*
IVANA MAHRÍKOVÁ
TOMÁŠ GIBALA
Department of Sanitary & Environmental Engineering,
Slovak University of Technology Bratislava,
Radlinského 11, 813 68 Bratislava, Slovak Republic

Abstract. The design of water line structures such as pipelines is a very complicated process since it depends on an iteration method which is not strongly mathematical. The main step depends on the designer who drives all the processes. This system of design produces many mistakes because the designer cannot consider all the objectives of the design. The design of pipelines requires an expert system. So if we want to produce the perfect plan of structures, which considers both technical and economical approaches, we have to use the expert knowledge of the designer in conjunction with maximum use of computer systems. This article concerns the development of a system for the complex design of pipelines. The software is called SEWDES and it was developed for designers to very quickly and perfectly design situations, develops longitudinal profiles, performs hydraulic calculations, and creates database statistics of the network. The outputs of the system are in DXF format, compatible with ACAD systems and expert hydrodynamic modeling systems such as MOUSE.

Keywords: Pipeline, design, runoff, water, sewage system, network, simulation model

1. History of pipeline design in Slovakia

The methodology for pipeline design in Slovakia was developed more than fifty years ago. The method was customized for handling both computations and drawings. This method was very slow, not very accurate and produced many

**To whom correspondence should be addressed. Stefan Stanko, Department of Sanitary & Environmental Engineering, Slovak University of Technology Bratislava, Radlinského 11, 813 68 Bratislava, Slovak Republic. e-mail: Stefan.Stanko@stuba.sk*

J. Pollert and B. Dedus (eds.), Security of Water Supply Systems: From Source to Tap, 137–145.

mistakes caused by the designer. The relation calculation and drawing technique was very heavy-handed. The designer did not have enough time for making the "approximation" between the mathematical design and drawings which were used as plans for building the structures.

The 1970s were the first experience with the aim of using computers in the design process. At first only the elementary tasks were solved by the computer and graphical outputs were only a dream of the designers. The computer hardware didn't allow the production of drawings. The end of the 1970s promised a better expectation for this dream. This period of the computer age was very surprising for engineers, but it was only the beginning. Hardware and software were lacking for two reasons: (1) political preference towards applications for the war industry and (2) the level of hardware development was low.

The expansion of software development continued into the 1980s, when the computer industry developed microprocessors which yielded the first use of personal computers. This allowed for both computations and the graphical display of outputs. It was a period when some young engineers had the opportunity to use the software at the office in the absence of a big computer center.

1.1. NEW PERIOD OF DESIGN

The 1990s brought enormous growth of computer hardware and software. This included new, not very expensive, peripheral equipment such as plotters; new "quick" processors; the minimization of computers and the maximization of operational memory; permanent memory and the quick graphical interface. Graphical cards were a major step for software developers. This allowed civil engineers to produce software with a new approach.

Various software programs were developed in this period, including video games but powerful tools for design structures in the civil engineering branch were also produced. One of these tools was a system for pipeline design developed at the Slovak Technical University – Faculty of Civil Engineering. The most recent beta version was developed in 2005. This tool is a complex system for defining inputs, performing calculations, easily changing the net topology and consequently, quickly producing the drawings as plans for building the line structures.

2. Methodology of design

The methodology for the complex approach to designing pipelines evolved from an older system, which primarily worked only with numbers and lacked a graphical display. This system, developed in 1989, defined the basic structures of the net data. Data inputs were primarily defined in this software. The system was written in FORTRAN, required a great deal of data inputs and was oriented toward scientific calculations. With the advent of Personal Computers, new opportunities appeared. In 1990, the system was enriched with an input module and a user-friendly module for producing the graphical view of the situation and consequently a view of longitudinal profiles on the screen. This allowed for better checking of the design. The programmed language was PASCAL. The operating system was PC DOS.

In 2004, many improvements were added to the software. The graphical user interface allowed for the definition of the required output as well as the longitudinal profiles. The main drawback was the precision and memory restriction of the numbers of elements (only 700 points were allowed).

The big change occurred in 2005. This change included a new software processor and a new approach for data definition.

2.1. INPUT DEFINITION

In the original system, data input resulted in many errors because the designer was requested to manually input the data. The new system provides two basic approaches for data input: (1) input from a CAD system, in this case the ACAD, and (2) an easy to use data definition as part of the system environment.

2.2. INPUTS FROM THE CAD SYSTEM

If the designer decides to use the CAD system, which is usually available, the system of data inputs is based on the method: “draw and XY log”. This method is briefly described as: To draw the points of the nodes, connect the lines between nodes and with the special software, very accurately record the XY data of the nodes for the next processing step.

The module for processing the XY data is called DIGITAL. It produces precise input for the design system environment. It contains the XY data, lengths between nodes, and the system of net topology contained in the net matrix. It automatically names the branches of the net. The system omits the level data, because it’s impossible to obtain it from the 2D plans. The designer has to supply the levels.

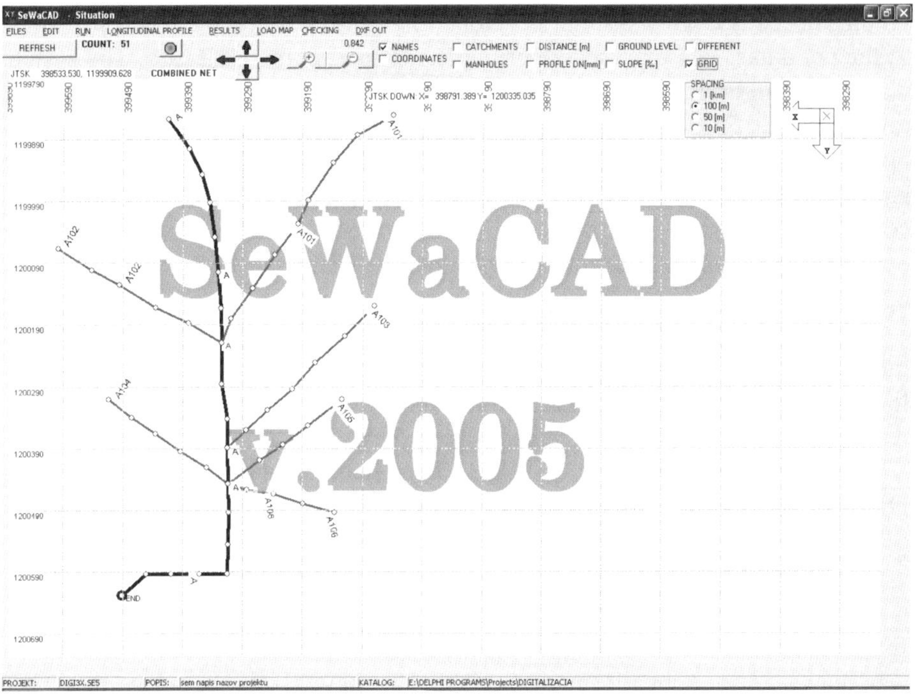

Figure 1. The output from ACAD environment, transformed through the digital routine

2.3. MANUAL INPUTS IN THE SOFTWARE ENVIRONMENT

Manual inputs provide for editing the data directly in the software environment. Manual inputs allow for the expansion of the net upstream and downstream. It also allows for the definition of length if exact distances need to be preserved. If length is omitted, the designer can define the exact placement of the new node. After definition of the new node, editing of this node is allowed.

Through a point and click interface a node can be easily found in the database. A table of node attributes can be displayed and edited. The data is updated after clicking on the change button.

Moving the node to a new position is possible in two ways: (1) entering the X, Y coordinates, or (2) moving the node with the mouse to the new position. This process insures the auto-recalculation of distances.

The system allows for the quick renaming of entire branches by finding the branch and defining the new name using the "combo-box" routine.

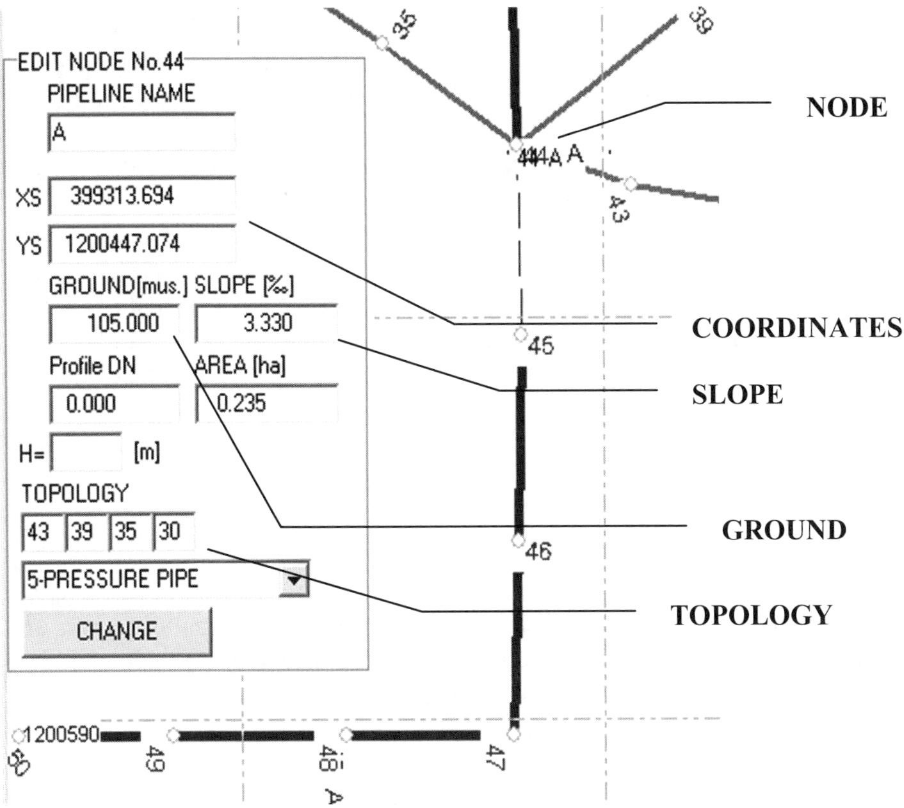

Figure 2. The editing table of the node

The manual input process is performed using the graphical user interface. The joining of the database with the graphical display allows for visual checking of the net by the designer and a very quick orientation of the design.

The difference between the CAD and manual input is as follows. The CAD input globally defines a new net with many nodes. This allows the designer to define not only one node at a time, but the entire net. The main advantage is a very quick definition of the net, with distance calculations and automatic naming of the branches. On the hand, the manual input allows the user to perform individual editing of nodes defined in the system and also to append new nodes with automatic insertion of system attributes.

2.4. DATA CHECKING

The software includes a module for data checking. This allows for a logical system check and prevents mistakes which have a negative influence on the hydrotechnic calculations. This is a very important module because in large nets, it is impossible for the user to manually check the data. This checking process concerns mainly net topology. The check is enabled by pressing a key and very quickly the user can watch the results of the checking process on the screen. The system shows the nodes where problems were detected. The user can very easy click on a node that contains a mistake and correct the errors.

A work session is ended by saving the data. The system contains a green - red icon which indicates whether the changed data was saved or not. The green color means the data was saved. The red color indicates that the data is still in computer's buffer. It enables the user to decide whether changes will be saved to the input file or not. This approach was chosen because the user may not always want to save the changes. It can also prevent the data from being damaged.

3. Slope generator

The design of longitudinal profiles is very challenging. The user needs to spend a lot of time in the manual design of the profiles. This challenge was the primary motivation for creating the slope generator module. This module allows, in a short period of time, the design of slopes for the entire net. It also allows for changes to parts of the net.

The slope design system allows the user to design the slopes by several approaches:

1. optimal design by computer
2. landscape replication
3. defining designed slope
4. function for slope generation, based on the idea "bottom to bottom"

The main function is the optimal design by computer. This procedure follows the minimum values of slopes defined by the profiles and takes into consideration the ground. It attempts to keep the slope at optimal depth under the terrain. The minimal slopes define the Slovak standard.

4. Pyrotechnic calculation

The hydrotechnic calculation is the process whereby the input data are transformed to a binary file which contains the hydrotechnic results. This data is displayed in graphical form on the screen.

The dimensions of the profiles are shown in various colors. This makes it possible to easily see the dimensions of the profiles on the entire net. The hydrotechnics calculation module allows for changing the type of the net and the amount of population and water per inhabitant per day. The calculation is easy and immediately realized using only the input file. The user can view the maximum flow of the net without using the module for presentation of results.

This system for the hydrotechnic calculation was the main module, but with the increase in software development it became a very simple module with the main work consisting of data editing. On the other hand, the calculation module contains verification techniques for profile design which is very important.

4.1. NODE NUMBERING

The Slovak system of numbering the nodes at the end is different from the numbering of inputs. The software uses a module which allowed for numbering of the nodes by the system. There are two approaches – full automatic, or manual driving.

4.2. LONGITUDINAL PROFILE - CALCULATION

The system uses two basic steps to generate longitudinal profiles. After using the hydrotechnic calculation and the numbering module, the user can choose from the menu, the module for computing longitudinal profiles. This module computes the profiles and each of them is saved in a separate file which is input for other modules.

The calculation of longitudinal profiles allows the user to compute the profiles with several options: bottom to bottom, axis to axis, top to top and water level to water level.

4.3. LONGITUDINAL PROFILE

After generation of the longitudinal profiles, the user can view these profiles on the screen. The system of viewing was designed with the idea to view the maximum amount of the design with minimum use of the keyboard and mouse. Thus, the user can see all the longitudinal profiles only by using the mouse wheel. The drawing details are viewed by zooming-in and zooming-out, by clicking the left or right mouse buttons, or using the buttons on the screen. This allows the user to follow the changes in the system.

If the user wants to make changes in the profiles, the system allows for the use of the resident editor for defining depths and switching the view to allow changing the details of the nodes.

The view of longitudinal profiles shows not only those profiles which are suitable as plan drawings but also for the creation of the optimal design. For the final drawings, the system contains a composite module for the generation of profiles in DXF form.

4.4. SITUATION

Situation is the basic idea in the system design. Creating a situation means defining the input which is dependent on ground level, terrain topology and local building plans. The user can watch the situation at each time in the design. Drawing and modifying the situation is possible by changing coordinates, appending the nodes or removing the nodes. It is also possible to display the city map which allows accurate orientation with for node manipulation.

The final outputs of the situation are passed to the module to export the situation to a DXF file.

4.5. CREATING DRAWINGS

There are two main modules for creating drawings. The first concerns the longitudinal profile and the second concerns the situation to generate. Both of them produce DXF files which can be read by various CAD software.

To generate longitudinal profiles, the user need only show the desired profiles to draw, define the files with the materials to be used and specify the output scale. This process automatically generates a DXF file.

The generation of the DXF file is the final procedure which the user employs after the optimization of the net system. The output is a drawing and this can be changed only in some CAD systems, e.g. ACAD.

The module concerning situation generation is in principle different from the longitudinal profiles generator, but the outputs are again in the DXF format which can be additionally edited in the CAD system.

4.6. MODULE FOR DATA TRANSFER TO MOUSE SYSTEM

The advantage of this system is that it is modular which enables the transfer of data to the format for the MOUSE 2003 hydrodynamical simulation model. This makes the system more operational and useful. The main advantage is the ability to perform a quick design and consequently perform modeling in the MOUSE system. Another advantage is that it is a useful tool to create input data to the MOUSE model.

Experiences with using this module show that the design by this software provides good correspondence with the calculations from MOUSE results.

5. Conclusion

The system exists in both Slovak and English versions. The system was developed using the DELPHI programming language. DELPHI is a very sophisticated programming environment and allows for the development of applications that are user friendly and with maximum capacity for expansion.

The system is designed for the very quick effective design of pipelines with dimensions and allows fast output data transfer for drawings of longitudinal profiles and various situations which are used as plans for building the pipelines.

An advantage of the system is its interface to the MOUSE expert system, consequently the system of pipeline design acts as an input editor for a hydrodynamic simulation model.

The main purpose of the system was to rapidly create the pipeline plans for the villages and cities in the Slovak republic.

Users of this system need some experiences with CAD and knowledge of design procedures so the system can be effectively operated.

Acknowledgement

The Research Grant VEGA 1/0310/03 held by the Department of Sanitary Engineering Faculty of Civil Engineering, Slovak University of Technology Bratislava has supported this paper

References

ČSN 73 6701 Sewerage systems & household connections.

Urcikán, P.: Analyze and using of rainfall data and hydrotechnical calculations of sewerage systems. Academic dissertation, STU Bratislava, 1987.

Stanko, Š.: Implementation computer systems to the sewerage system design – software SEWDES. Seminar Tempus-Phare, Faculty of Civil Engineering, 1999.

Urcikán, P., Imriška, L.: Sewage systems and waste water treatment – tables for calculations of pipeline dimensions. ALFA/SNTL, 1986.

Act No.442/2002 Dig. on public water and sewerage systems.

ČSVTS, Provoz a údržba stokových sítí, Sborník přednášek, 1987.

Stanko, Š.: User Manual - SEWDES v4.0, Department of Sanitary Engineering, Faculty of civil Engineering, Slovak University of Technology, 2004.

WATER RESOURCES OF THE CRISURI RIVER BASIN – QUALITY & SECURITY

OCTAVIAN STRENG*, CEZAR MORAR
Crisuri Rivers Authority, National Romanian Water Administration, Ion Bogdan Str. 35, Oradea 410 124, Romania

Abstract. The paper has two major parts. The first part is about the water resources in the Crisuri River Basin. General information about the hydrographical network and a detailed analysis of water resources throughout the basin are provided. The present water resources, their distribution in time and space, and resources for future needs are described. The population served by centralized water supply systems, waste water systems and treatment water stations is presented and the future trends in the European Union integration context are discussed. The second part of the paper is about the water supply system in the City of Oradea, the most important town in the basin. The catchments, treatment and pumping of water in the system, transportation and delivery of the drinking water network are issues are discussed. The general city plan of the water delivery network, its present status and future trends are shown. How the drinking water quality is monitored and the security measures taken are shown in the last section of the second part. The conclusions present information about the sustainable management of water resources and improvement of water resource quality.

Keywords: water resources, water supply system, management, EO data, GIS, Crisul Rivers Basin, The City of Oradea, drinking water

**To whom correspondence should be addressed. Octavian Streng, Crisuri Rivers Authority, National Romanian Water Administration, Ion Bogdan Str. 35, Oradea 410 124, Romania*
E-mail octavian.streng@dac.rowater.ro

J. Pollert and B. Dedus (eds.), Security of Water Supply Systems: From Source to Tap, 147–154.

1. Water resources of the Crisuri River Basin

1.1. INTRODUCTION

The Crisuri River Basin is located in north-western Romania and eastern Hungary with a total surface area of 27 500 km^2 (14 860 km^2 on Romanian side ~ 54%) This represents 6.3 % of Romania's total area). The main rivers (Barcau, Crisul Repede, Crisul Negru and Crisul Alb) cross from the north to the south of the Romanian - Hungarian border, merging two by two on the Hungarian side in Crisul Dublu (Double), then Crisul Triplu (Triple). The discharge is to the Tisa River near the City of Csongrad.

The Romanian side of the basin contains 365 water courses; the total hydrographical network length is 5785 (representing 7.3% of the total Romanian Hydrographical Network) –and the density is 0.39 km/km^2.

Throughout the basin there are 3 geomorphologic areas: mountains (about 38%), hills (20%) and fields (42%) arrayed in steps from east towards west.

Figure 1. Crisuri River Basin location in Romania

1.2. WATER RESOURCES

The main rivers in the Crisuri Hydrographical Basin are the Barcau (2 025 km^2), Crisul Repede (2 973 km^2, Crisul Negru (4 230 km^2), Crisul Alb (4 240 km^2). Crisuri River Basin water resources are sufficient for local needs; though they are non-uniformly distributed in time and space. The main water sources in the Crisuri River Basin are ssurface waters, represented by rivers and storage lakes (reservoirs) and groundwater, represented by phreatic and deep waters. The total theoretical water resources in the basin are about 2937, 4 million m3 (surface waters and groundwater) but only 25.3 % are technically useable (745 million m3 out of which 395 million m3 originate from rivers and storage lakes and 350 million m3 from groundwater). From all available resources, only 95 mil. m3 (13%) are used, so there should be sufficient resources for future needs.

1.3. CURRENT SITUATION

The population in the Romanian Crisuri Basin territory is 875,366 people (2002) out of which 41% lives in urban areas and 59% in rural areas. There are 4 cities, 13 towns and 177 villages. The main city is Oradea – the capital of Bihor County –with a population of 220,000.

The total number of water supply systems is 54 with one of these systems providing service for more than 200.000 people (the City of Oradea). Water supply services are performed by 4 private or mixed companies, 2 autonomous authorities, and 48 public services. Water supply is provided by centralized systems for 41, 5% of inhabitants (83, 6% in urban areas and 16, 4% in rural areas).

Table 1. The number and percent of inhabitants connected to water supply waste water systems and water treatment services (in 2002)

People connected to centralized water supply systems (No. of people and %)			People connected to waste water systems (No. of people and %)			People connected to treatment water stations (No. of people and %)		
Total	Urban	Rural	Total	Urban	Rural	Total	Urban	Rural
366 814	306 664	60 150	255 884	251 234	4 650	242 778	238 663	4 115
41.5%	83.6%	16.4%	29.2%	98.2%	1.8%	27.7%	98.3%	1.7%

1.4. TRENDS

Romania's evolution, the civilization level, and population living standards depend significantly on the development of water supply and waste water systems. In its efforts for integration into the EU in December 2004, Romania closed *EUROPEAN UNION* negotiations at *22nd Chapter – ENVIRONMENT.* By these negotiations E.U. accorded Romania transition time till 2018, when all the inhabitants must be connected to water supply, and waste water collection and treatment systems.

2. Water supply system in the City of Oradea

2.1. THE SUPPLY, TREATMENT & PUMPING OF WATER

In the city of Oradea, the raw water supply comes through five waterworks placed on the two banks of the Crisul Repede River and located in the north-eastern part of the city, having a total discharge of 2100 l /s.

The raw water is collected either directly from the river or from the underground layer through system drains

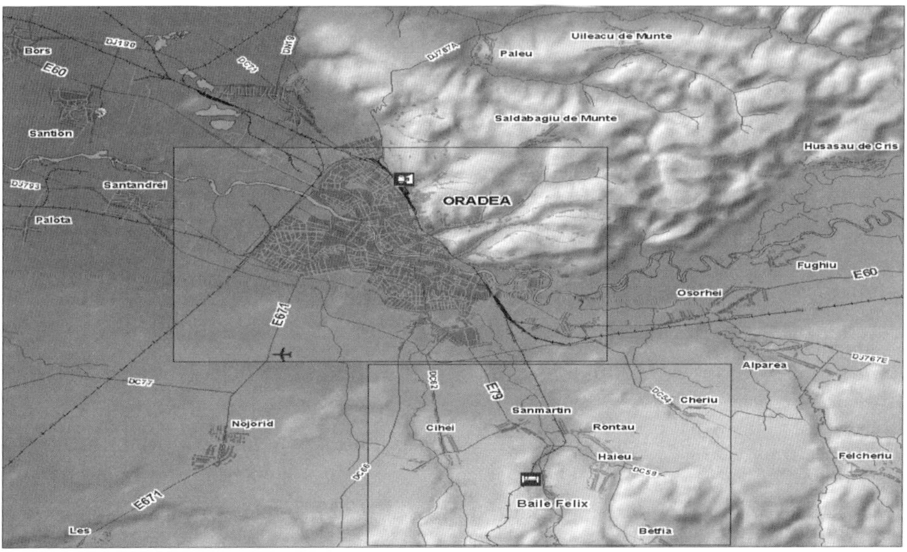

Figure 2. The City of Oradea

Because of the variation in the Crisul Repede discharge (from 1 m3/s to 600 m3/s), in 1974 the Lesu permanent water supply reservoir was constructed upstream on the river, in the mountains. Lesu reservoir has 28 mil. m3 capacity and the purpose of this storage are as a drinking water reservoir.

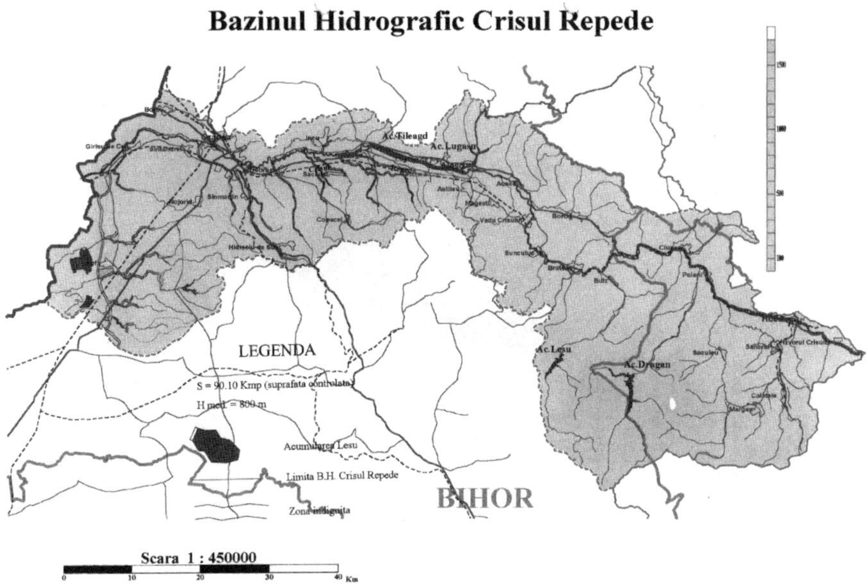

Figure 3. Crisul Repede River Hydrographical Basin

Therefore, in the waterworks area, a maximum insured flow was secured to produce drinking water. The complex hydro-technical works on the Crisul Repede River and its main tributaries started after 1974 - for power, flood mitigation and water supply purposes. To support these purposes, three big reservoirs were built: Drăgan – Floroiu (on Drăgan Valley with a maximum volume of 120 mil. m^3), Lugaş (on Crisul Repede River with 63.5 mil. m^3 volume), and Tileagd (on Crisul Repede with 53 mil. m^3 volume).

Upstream of the water-works area a compensation reservoir for downstream discharge control was built ensuring between 6 and 40 m^3/s flow for water supply.

The most important raw water source is from underground. For underground water enrichment, 23 basins are used (15 on the right and 8 on the left bank of the river). The basins are supplied with water from the Crisul Repede River, through pipe adductions.

Even if the existing technology allows the use of surface water directly from the Crisul Repede River, with a proper treatment, this solution is a back up solution with underground water being preferred because of its better quality and its lower cost requiring only treatment with chlorine. The whole catchments system, water adductions, enrichment basins, infiltration fields, water works are situated in a severe protection area with a total surface of 280 ha.

2.2. TRANSPORTION AND DELIVERY DRINKING WATER NETWORK

The city water supply system delivers water through principal or secondary pipes. The principal network supplies water to companies, institutions, to two-story buildings, and to the hydrosphere pumps (83 re-pumping stations, out of which 77 are modernized.

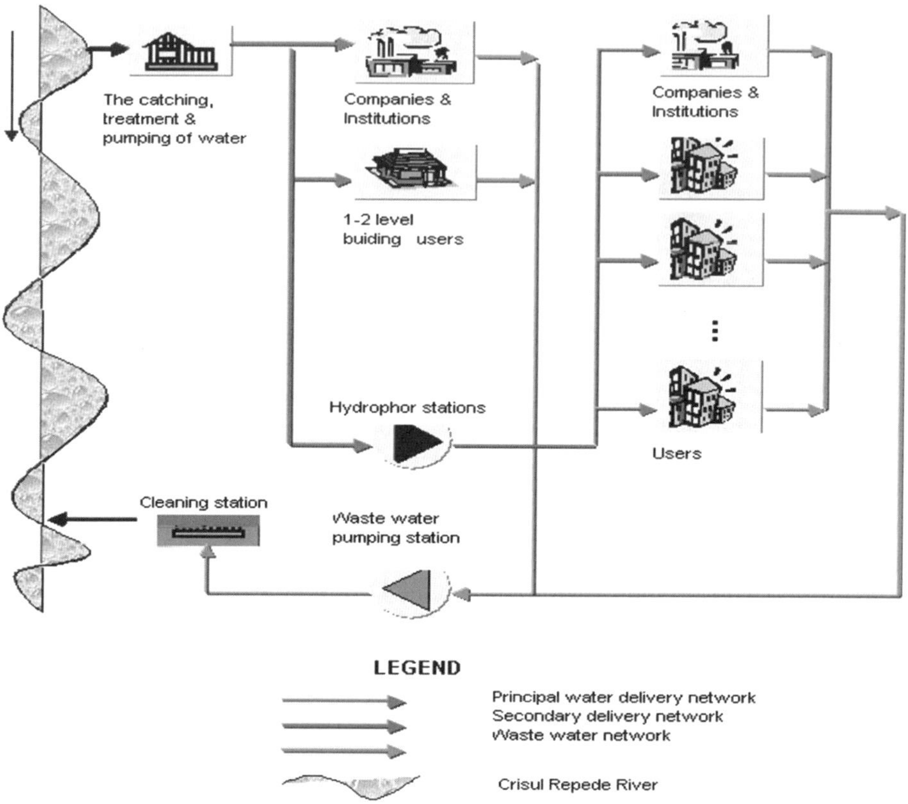

Figure 4. A schematic of water supply system in the city of Oradea

There are 142 km of secondary pipes in the water supply network, all of which are fully rehabilitated (made of polyethylene) and serve especially the taller buildings supplied with water by the hydrophor pumps.

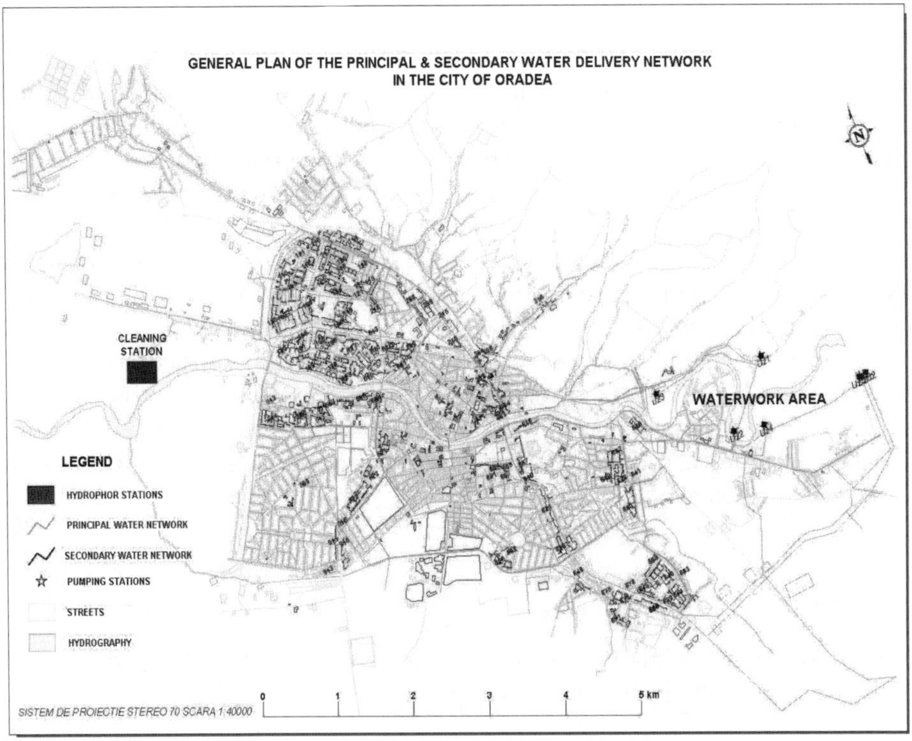

Figure 5. General plan of the water delivery network in Oradea

2.3. MONITORING DRINKING WATER QUALITY

Drinking water quality is analyzed in two laboratories, which perform daily analysis of microbiological, organoleptical, physical, and chemical indicators. Both raw water and chlorine pumped water in the delivery system are monitored. Special securities measures are taken when the water quality is not adequate correspond for different reasons, primarily because of elevated turbidity. In these situations the waterworks close the water supply system, until the quality parameters return to acceptable standards. System security at the water-works is provided for a severe protection area with permanent monitoring by its employees.

According to the law, water companies must prepare a yearly Monitoring Plan for the treatment stations, reservoirs and supply areas. This action is done together with the Local County Health Authorities.

3. Conclusions

Regarding the sustainable management of water resources, the concept of integrated water management involves both the use of water resources and the protection of natural ecosystems. Thus, the following objectives are taken into consideration: ensuring continuous water supply for all users and especially for the population, creating new water sources, especially through integration with complex uses in regions with water deficits creating separate water supply systems for the population and for industry, saving water and reducing losses in water distribution systems.

Improvement of water resources quality depend upon acquiring newer technologies for the production processes and using clean, non-polluting technologies, developing new wastewater treatment plants and modernizing existing facilities and implementing means to prevent, limit and diminish the effects of accidental pollution.

References

Daniela Teodorescu, *Water Resources. European Legislation*,*H*G*A, Bucuresti, 2002.
Constantin Trufas, *Water Quality*, Agora, Calarasi, 2003.
Crisuri Rivers Authority, http://www.apecrisuri.ro
Oradea Waste Water Agency, http://www.apacanal.ro

SECURITY OF WATER SUPPLY AND SEWERAGE SYSTEMS IN SLOVAKIA - PRESENT STATE

K. TÓTHOVÁ* , I. MAHRÍKOVÁ
Department of Sanitary Engineering, Slovak University of Technology Bratislava,
Radlinského 11, 813 68 Bratislava, Email:

Abstract. Water supplies are among the most critical infrastructures, such that their paralysis or destruction could weaken a country's defensive or economic safety. A great deal of attention is being given to the preparation of preventive measures, safety plans and risk analyses of water supply vulnerability because they provide both basic a human need and are important for the health of towns and villages. There is a lack of complex emergency response plans for water companies in the Slovak Republic. Water companies have developed records for emergency and alternative water supply within frameworks of operative orders for water piping as stipulated by the Slovak Act on public water and sewage systems, or they have developed internal regulations on emergency supply solutions defining regulatory stages and responsibility of the operational staff. Emergency response plans are missing in such materials as well as the provisions on awareness of a crisis situation. Lately, much attention has been given to this issue and authorities have started to require the development of vulnerability assessment and emergency plans for water supplies.

Keywords: Water supply and sewerage systems in Slovakia, legislation about water supply in crises situation, regulatory measures, emergency and alternative water supply.

* *To whom correspondence should be addressed. Katarina Tothova, Department of Sanitary Engineering, Faculty of Civil Engineeering, Slovak University of Technology, Radlinskeho 11, 81368 Bratislava, Slovakia; katarina.tothova@stuba.sk*

J. Pollert and B. Dedus (eds.), Security of Water Supply Systems: From Source to Tap, 155–168.

1. Introduction

As a result of increasing interconnection and information the world has never been so vulnerable. New and changing vulnerabilities have arisen from a more open global society, complex technological systems, greater dependency on electronic information and communication systems, integrated systems of food production and supply, and interconnected large transport systems. Inactivation of a few pivotal major points and functions for a long time – irrespective of whether it is a result of a terrorist attack, sabotage or a technical failure – could bring a large breakdown.

The paralysis or destruction of the most critical infrastructures could weaken the defensive or economic safety of a country. These infrastructures include telecommunication, power and transport networks, banking and financial services, state and emergency services and, last but not least, water supply systems. For the above reasons, a great deal of attention is being given to the preparation of preventive measures, safety plans and risk analyses of water supply and sewerage system vulnerability because they provide both a basic human need and are important for the health of towns and villages.

2. Conception of inner safety in the Slovak republic (SR)

A state safety environment is currently characterized by overcoming the threat of a global military confrontation of bipolar groups. Furthermore, the possibility of a threat, crisis or low intensity conflict, that could lead to a joint combination of more dangerous threats is of growing concern. The safety environment is further characterized by a large number of diffrent threats coming from diffrent directions and geographical locations. Outside acting threats can be combined with threats inside the state. An open economy provides the SR with the opportunity for wider political and economical cooperation within the European area. The SR gradually strenghtens democratic groups, improves the functioning mechanisms of the legal state, creates conditions to form a civil society, and adopts systematic measures for the completion of a fundamental reform and transformation of its social, economical and environmental sectors. It is a substantial part of an international community and an active subject of international relations.

Being attached to the basic principles and values embodied in the Washington agreement and being prepared to contribute to the stability and security in crisis and conflict situations, the SR fulfilled its efforts by joining NATO. The safety environment in Middle Europe is stabilized but it seems to be still more and more affected by outside factors that are out of individual states´control and that are not respectful of international liabilities and responsibilities. Resulting from a whole world trend of economic, financial, cultural and information globalization, these factors lead to a rise of non-military safety threats having an asymmetrical character quite often. Dependency on the preservation of a needed safety standard of the state and on effective common activities of an international community against sources of threats is increasing not only when solving accute but also chronic crises within an international framework.

3. Public water supply and waste water collection

3.1. WATER SUPPLY CONNECTION RATIO AND WATER CONSUMPTION

The total number of inhabitants connected to the public water supply network in 2003 was about 84.3 % (4.54 million inhabitants). Development of the public water supply from year 1990 is shown in figure 1.

Despite the rising tendency in the number of supplied inhabitants, a decrease in drinking water consumption and household specific water demand remained. In the course of the years 1990 – 2003, specific water consumption in households dropped from the level of 195 l/inh/day to 109.2 l/inh/day. This phenomenon is caused partially by decreasing industrial production, but also by the gradual decrease in public water consumption due to increasing drinking water tariffs. The specific demand for drinking water has decreased since 1990 by almost 40 %. Compared to EU countries, which consume approximately 145 l/cap/day, the specific water demand in Slovakia is low. The consumption of drinking water is falling well below the average, leaving us dangerously close to approaching the minimal hygienic standards. This trend also applies to other water use categories, since the demand of industrial and other users – compared to 1990 – has decreased by 30 %. The amount of unaccounted-for-water (UFW) is at the level of 30 % of the produced water; out of this amount more than 79 % covers losses in the piping network (24.0 %).

3.2. REGIONAL (LONG-DISTRIBUTION) WATER SUPPLY SYSTEMS

Non-uniform occurrence of water in time and space in different regions of the country has led in the past to development of large-scale water supply systems. Due to this, there are 47 regional water supply networks currently supplying some 70 % of the inhabitants with good quality drinking water (i.e. 80% of all produced water in the country). Interconnections of the regional systems have resulted in the formation of supra-regional systems, such as the Western Slovakian, Central Slovakian or Eastern Slovakian Water Supply Systems, which are based on combined use of several large ground water resources, and other emerging systems. It is anticipated, that future development in this area will focus on utilization of existing and developed capacities over exploitation of new resources.

3.3. FUTURE DEVELOPMENT OF WATER SUPPLY

The priority in the field of water supply for the future is to increase the proportion of inhabitants connected to public water supply systems to the level of the EU countries and to achieve a 97% connection ratio. The overall budget needed to secure this goal is estimated at SKK 70 billion (EUR 1.7 billion). It is clear that it can only be achieved by foreign assistance, i.e. through the pre-accession instruments or via the Cohesion fund and structural funds.

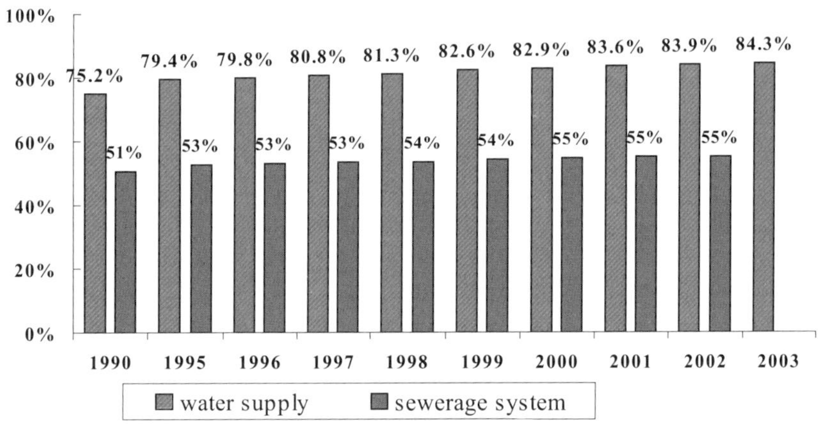

Figure 1. Comparison of the amount of population supplied from public water and the amount of population connected to public sewerage systems in percent (%)

3.4. WASTE WATER COLLECTION AND TREATMENT

Present stage

From the 2,883 municipalities in Slovakia only 557 of them are connected to public sewerage systems, i.e. 19.3 %. The number of municipalities connected to public sewage systems with following treatment in WWTPs is 492, i.e. 17.1%.

The Common Position of the European Union in Chapter 22 Environment, approved by the EC and SR at the end of 2001, appreciates in part Water

Quality and the fact that the Slovak Republic will designate the entire territory as sensitive in the sense of Article 5 of the Directive 91/271/EEC. The EU notes that advanced treatment will be provided for each treatment plant above a population equivalent of 10 000, i.e. that nitrogen removal will be ensured in accordance with Annex I of the Directive and that phosphorus removal will be ensured where specific circumstances of waters so require. This agreement is implemented in Gov. Regulation 491/2002.

General surface water quality objectives and effluent standards in Slovakia are set in Government Regulation No. 491/2002 and applied by the local Environmental Agencies when issuing discharge permits.

The Regulation was prepared with the aim to correspond with European legislation, especially with the Council Directive 91/271/EEC. This Directive is a follow up to the new Slovak Water Law. It represents a fusion of ambient water quality standards and end-of-pipe effluent standards common in European countries:

- effluent standards are defined both for municipal WWTPs and selected industrial waste waters (in accordance with the UWWT Directive),
- ambient water quality standards are defined for the receiving water (see table 1).

Water authorities have the right to set stricter effluent standards calculated from the "mixing equation" based on the ambient standards.

Table 2. Transitional measures required under Slovak Directive 91/271/EEC

Agglomeration (AGM)	Number of AGMs	Proposed implementation	Trans. period (from 1.1.2004)
Sewerage systems			
AGMs with >10,000 PE	18	2010	6 years
AGMs with 2,000 –10,000 PE	409	2015	11 years
Waste water treatment			
AGMs with >10,000 PE	90	2010	6 years
AGMs with 2,000 – 10,000 PE	439	2015	11 years

Future development

Table 2 presents the requested transitional measures that have to be completed after the affiliation of Slovakia to the EU as stated in Chapter 22 of the Common Position of the EU.

4. Water supply in times of crises in the Slovak republic – legislative basis

Act No.442/2002 Dig. on public water and sewage systems is a legislative framework of a public supply and waste water collection. According to this act, a public water supply shall be understood as a complex of structures and equipment serving public needs and allowing a mass water supply for inhabitants and other consumers. A public sewerage shall be understood as an operationally independent complex of structures and equipment serving public needs and facilitating a public derivation of sewage water allowing an environment-friendly entry, discharge and usually waste water treatment. A public water supply shall be understood as a water supply for more than 50 persons or a supply with an average daily consumption larger than 10 m^3. A public collection of a wastewater shall be understood as an entry, discharge and usually a waste water treatment from more than 50 persons or when an average daily waste water outflow exceeds 10 m^3.

Act No.442/2002 Dig. on public water and sewage systems speaks about smooth water supply and waste water collection and stipulates conditions for a water supply interruption or reduction because of an extraordinary situation. These situations include: water-service pipe or sewer pipe failure, if life, health, or property is directly threatened, in the case of planned repair, maintainance and revision works, in the case of water supply restrictions or water consumption control stages. Pending a drinking water shortage, a municipality, as public administration executive in the sector of public water pipes and sewerage, issues a statute of binding force on a form of an emergency water supply and waste water draining in terms of the Act on public water and sewage systems, and according to local conditions and statutes temporarily limits or surpresses use of the drinking water for other purposes than necessary for the drinking water supply in the municipality.

If an extraordinary situation leads to a water supply limitation or surpression, the operator must follow the Act of the National Council of the Slovak Republic No. 42/1994 Coll. on Civil Protection of Population as worded in later regulations and Act No. 387/2002 Coll on Governing the Country in Crisis Situations except in a Time of War or Time of War State. These acts regulate conditions for effective preservation of life, health, or property against results of extraordinary situations and the competence of the public authority bodies and municipalities, rights and duties of legal persons and physical persons for securing the civil protection of population.

According to the Act on civil protection, an extraordinary situation shall be understood as a condition caused by a natural disaster, average or catastrophe:

- *Natural disaster* is an extraordinary event when unwanted release of accumulated energies or material occures as a result of natural forces, when a dangerous materials or destructive factors could act with a negative effect on life, health, or property,
- *average* is an extraordinary event that causes a derogation from a stabilized operational state that leads to a leakage of dangerous materials, or to an action of other damaging factors with influence on life, health, or property,

- *catastrophe* is an extraordinary event of growing savage factors and their acccumulation resulting from a natural disaster and average.
- A time period of the presumed danger of extraordinary situation´s uprise or extention is called emergency.

A civil protection shall be understood as a system of tasks and measures aimed at the preservation of life, health, or property by means of an analysis and valuation of safety risks and dangers, as well as a specification of measures for decreasing risk and activities and measures on extraordinary situations to control consequences.

A threat period or a period of exceptional situation consecutive effects on life, health, or property, that is declared according to this law regulation and during which measures aimed to decrease risks or measures and activities aimed on control of the extraordinary situation effects are executed, is called an extraordinary situation.

An emergency supply is closely connected with an emergency state that is one of the tasks of a civil protection. Within its framework a minimal catering, minimal water delivery and securing of basic needs of the people affected by the extraordinary situation is being realized within the bounds of existing survival conditions.

An obligation to secure an emergency supply arises not only in the case of extraordinary situations at peacetime but also when the extraordinary situation arises during a time of war (autrefois state defence emergency). In that case the water supply falls within the cognizance of Act No.411/2002 Coll on Economic mibilization and on a modification of Act of National Council SR No. 274/1993 on the Scope of Competence of Relevant Bodies in Consumers Protection Matters as worded in later regulations (Act No. 511/2002 and 622/2003 Coll).

An economic mobilization creates conditions for securing population essential survival needs and for securing activities of armed forces, armed security corps and other forces during the crisis situation.

Organization of drinking water supply at crisis situations in peace and war conditions, at extraordinary situations as a natural disaster, average, catastropohe, epidemic and terorism is governed by a procedure of the Ministry of Interior SR by which details on the emergency water supply at crisis situation are stated. This regulation is currently prepared for adoption by the National Council and is to enter into force on 1st July 2005.

According to the mentioned regulation, the preparation of a water supply at a crisis situation is given by planning and adopting preventive measures to secure a potable water supply for population, subjects of the economic mobilization, armed forces, armed security and other corps. Preventive measures shall be elaborated on the basis of two issues:

- current situation analysis of the water supply
- evaluation of security risks and threats.

The current situation analysis of the water supply in the area of valuation includes:

1. characteristics of the water supply area where physico-geographical, demographical circumstances and a form of the drinking water supply,
2. analyses of a current state of the water supply based on data on the water consumption for inhabitants, industry, agriculture, amenities and on a basis of data on a water supply system,
3. evaluation of the influence of an energy system on the water supply in terms of the drinking water production and its transport to consumers.

Evaluation of security risks and threats includes:

a) possible sources of a threat or a contamination of water resources, objects and equipment of a public pipeline
b) vulnerability of used water resources and waterworks objects to supposed security risks and threats,
c) possibilities of disabling key water resources and waterworks objects and a proposal of an emergency drinking water supply after their disabling. This proposal is prepared for different scenarios.
d) possibility of energy system desintegration and a proposal for an emergency power supply.

The safety risk analysis and valuation is executed by the subject securing the water supply. In times of a crisis situation this subject is designated by the mobilization act. The Municipality is the designated subject in the field of the water supply and waste water collection. Safety risk analysis and valuation serves as a basis for elaboration of a crisis plan that defines the water resource used for the emergency supply in crisis situations.

On the basis of the safety risk and threats analysis and valuation, regulatory measures for the water supply are ordered implying a standard continual water supply. Regulatory measures are secured by regulatory stages 1 – 3.

Regulatory stage No.1 is ordered if the loss in water resources′ efficiency is more than 15 % , or when the water resources are disabled and the water piping can secure the drinking water supply in lower quantity up to 85% of the current water demand. By the regulatory stage No.1, using drinking water from the public water network for other purposes e.g. garden , playgrounds or public areas irrigation, or car washing is restricted.

Regulatory stage No.2 is ordered if the loss in water resources′ efficiency is more than 30 % , or when the water resources are disabled and the water piping can secure the drinking water supply in lower quantity up to 70% of the current water demand. In addition to measures stated by regulatory stage no.1, regulatory stage no. 2 includes measures for a suspension of drinking water supply for water heating and restriction in drinking water supply for major consumers.

Regulatory stage No3 is ordered if the loss in water resources′ efficiency is more than 70 %, or when the key water resources are disabled or when fundamental sections of a water piping public network are interrupted and the water piping can secure the drinking water supply in lower quantity up to 70% of the current water demand. In addition to measures stated by regulatory stages no.1 and 2, regulatory stage no. 3 states time of the water supply for the public water network .

For transport of lower water quantity by the public water piping, the volume of water supply is determined in multiple versions on the basis of the safety risk and threat results.

Respective versions of the water supply are based on:

a) water resources intended for the water supply allocation,
b) strength of the water source or a recommended water consumption from the water source,
c) technical state of the public water network after some water sources or waterwork objectss have been disabled,
d) energy intensiveness.

The means of the drinking water quality stabilization in the public water piping at lower supplies is stated for particular variants. Technical measures for the drinking water quality stabilization at regulated transport has to be handled at peacetime.

If the drinking water supply can not be secured by the public piping, it will be secured by an alternative supply. The alternative drinking water supply is a drinking water supply by means of water tanks or by other transport means. If the drinking water supply can not be secured by the alternative drinking water supply, it will be superceded by the emergency water supply.

The emergency water supply is a particular and limited form of the drinking water supply, by which only the minimal drinking water needs are covered by the drinking water supply. The drinking water supply is secured from public water piping units e.g. water treatment plants, water storage tanks, fire-hydrants or other sources designed for the emergency water supply.

Water resources for the purpose of the emergency water supply are ground water sources with a gravity water flow independent of energy resources and water treatment for water desinfection. If it is not possible because of natural conditions, surface water resources could be designed for the emergency water supply with a mobile water treatment facility.

The preferential water supply is for the population, subjects of the economic mobilization, armed forces, armed security corps and other corps.

The minimal emergency drinking water consumption is 15 l per person a day. In extraordinary severe conditions, the minimal drinking water consumption is 5 l per person a day for not longer than three consecutive days.

For subjects of economic mobilization securing the livestock breeding, the regulation states a minimal drinking water consumption as one head per day.

For subjects of economic mobilization in the section of civil amenities, the regulation states the emergency water consumption 30 l per bed daily or up to the technologic minimum.

The total emergency water consumption is stated as the sum of the emergency water consumption for population and the emergency water consumption for all subjects of the economic mobilization.

The mentioned Acts and Regulation Slovak Republic applies Slovak technical standard STN 75 5040 emergency water supply (1991). This standard applies the emergency supply at major industrial averages, natural disasters and war consequences. The standard defines technical and constructional provisions for use and security of water sources and structures for emergency supply.

4.1. DEMANDS ON PUBLIC FINANCES

Water supply in time of a crisis situation is necessary for the survival of the population and for military forces, armed security corps, and other corps involved in security actvities. In accordance with the respective acts on economic mobilization, Water Companies Inc. was designed as a subject of economic mobilization by a decision of the Ministry of Environment. The mentioned decision obliges water companies to execute economic mobilization measures and, at a crisis situation after declaration of the economic mobilization, to secure a drinking water production and supply at the extent of ordinary services pursuant to a capability and producability or pursuant to other requests of a public power for the solving the crisis situation.

Subjects of the economic mobilization already have to be prepared to secure this obligation in the peacetime. This is connected with a need of technical and material support, e.g. transport means (water tanks), mobile power plants, standby pumping devices, water treatment facilities, safeguarding means to guard and secure objects of special importance, etc.

This technical and material support will impact the state budget because these costs shall be covered by it as stated in the Regulation of Ministry of Finance SR 98/2003 Coll by which the details of the economy mobilization costs are governed.

Considering the seriousness and time consumption needed to secure this obligation, it will be necessary to release, every year from the state budget, financial means for subjects of the economic mobilization, designed by Ministry of Environment SR, according to legislative regulations of binding force on financing economic mobilization measures.

5. Emergency response plan - ERP

Regulation ME SR in preparation for a water supply in crisis situation, presumes that the subject of economic mobilization, that is the water company in a sector of water supply and waste water collection, has an elaborated emergency response plan. The obligation to elaborate this plan results from Act No. 411/2002 on Economic Mobilization., Other details and measures are governed by the Regulation of Ministry of Economy SR 119/2003 by which some of provisions of Act No. 414/2002 Coll on economic mobilization and on the obligation and execution of the Act of National Council SR No. 274/1993 on Scope of Competence of Relevant Bodies in Consumers Protection Matters.

The Regulation determines the structure, contents, extent and form of update and periodicity of data documents and information as well as the structure and contents of the emergency response plan.

The structure of the emergency response plan is designed in six chapters with logically structured information needed by the subject of economic mobilization to follow the economic mobilization measures. As a part of the emergency response plan structure an approval clause is regulated.

According to Regulation ME SR 119/2003 the emergency response plan consists of the following parts:

I. chapter - Methodology for performance of the subject to reach a state of readiness to execute measures of economic mobilization – step-by-step procedures for people fulfilling crisis management tasks.

II. chapter – Liabilities of the subject – information of the subject´s liability resulting from the Act on economic mobilization according to the particular states:

1. emergency state
2. exceptional state
3. war state
4. war

III. chapter – Valuation of conditions of the subject to fulfill its liability – data on its own material, energy, personnel, transport and other conditions to execute measures of economic mobilization and information on supply chain and state material reserves.

IV. chapter – Liability fulfillment needs – data on material, energy, personnel, transport and other needs.

V. chapter – Additional information on risk sources that may cause a crisis situation, on subject, on a place where papers needed to execute measures of economic mobilization are stored.

VI. chapter – Method of the economic mobilization measures execution – conclusions of chapters I. – V. sequenced according to an Act, Slovak Government decree, decision, order or legal regulation for emergency state, exceptional state, war state and war, and specification of procedures that shall be executed by the subject in order to fulfill a particular measure of the economic mobilization.

Approval clause – name, surname, function and signature of a statutory organization and a date of approval.

An obligation of the subjects′ statutory organizations to approve emergency response plans or data documents was given by a term of 15th January 2004.

6. Conclusion

There is a lack of complex emergency response plans for water companies on water supply and waste water collection in the SR. Water companies have elaborated records for emergency and alternative water supply and water collection within frameworks of operative orders for water piping and sewerage as stipulated by Act No. 442/2002 Coll. on public water and sewage systems, or they have elaborated internal regulations on emergency supply solutions defining regulatory stages and responsibility of the operational staff. Safety risks and system threats valuation is missing in such materials as well as as the provisions on awareness of a crisis situation. Lately, more and more attention is being given to this issue and it is only a matter of time when authorities start to require elaborated emergency plans for water supply and waste water collection.

Acknowledgement

The Research Grant VEGA 1/2137/05 and Grant VEGA 1/0310/03 held by the Department of Sanitary Engineering Faculty of Civil Engineering, Slovak University of Technology Bratislava has supported this paper.

References

Slovak Act No.442/2002 Dig. on public water and sewerage systems.

Regulation of Ministry of Slovak Republic Environment No 55/2004 stating requirements of operating instruction for public water piping and sewerage.

Slovak Act No. 387/2002 Coll on Governing the Country in Crisis Situations except the Time of War or Time of War State.

Slovak Act No.411/2002 Coll on Economic mibilization and on a modification of Act of National Council SR No. 274/1993 on the Scope of Competence of Relevant Bodies in Consumers Protection Matters as worded in later regulations (Act No. 511/2002 and 622/2003 Coll).

Regulation of Ministry of Slovak Republic Economy 119/2003 by which some of provisions of Slovak Act No. 414/2002 Coll . on economy mobilization and on an obrogation of Act of National Council SR No. 274/1993 on Scope of Competence of Relevant Bodies in Consumers Protection Matters are executed.

Regulation of Ministry of Slovak Republic Environment by which details on water supply in crisis situation are stated a document prepared for approval by Governmental Legislative Council SR.

RISK ANALYSIS OF WATER DISTRIBUTION SYSTEMS

LADISLAV TUHOVCAK*
JAN RUCKA
TOMAS JUHANAK
Institute of Municipal Water Management, Brno University of Technology, Zizkova 17, 602 00 Brno, Czech Republic

Abstract. The risk analysis represents a modern approach to determining the level of provision of drinking water supplies for the consumers and the safety of the whole drinking water supply system. The authors present the most frequently used methods of risk analyses of drinking water supply systems and address identification of qualitative as well as quantitative risks posed by the individual system components, the evaluation methods and interpretation of results. What is also described here are the basic principles of implementing the Hazard Analysis and Critical Control Points (HACCP) method and the possibility of transferring some experience and methods from other fields of industry, which might be more advanced in this area. The presented methods are demonstrated on a case study of a risk analysis concerning a selected section of a water supply system.

Keywords: risk analysis, hazard, HACCP, water supply, reliability, water quality

1. Introduction

The safety of drinking water depends on a number of factors, including the quality of source water, effectiveness of treatment and integrity of the water

* *To whom correspondence should be addressed: Ladislav Tuhovcak, Institute of Municipal Water Management Brno University of Technology, Zizkova 17, 602 00 Brno, Czech Republic, tuhovcak.l@fce.vutbr.cz*

J. Pollert and B. Dedus (eds.), Security of Water Supply Systems: From Source to Tap, 169–182.

distribution system (WDS) that transports the water to consumers. The current practice of designing and operating water systems very often employs methods of mathematical modelling. However, the risk theory is applied to the development and the use of the models rather rarely.

The risk analysis of water supply systems has only been used for a relatively short period abroad and the Czech Republic practitioners are just starting in this field, using experience, methods and findings from other branches of industry. Due to extreme floods affecting Central Europe in 1997 and 2002, greater attention is now being paid to the safety of drinking water supplies in the Czech Republic during these crises. Other major impulses for implementing the risk analysis for the design, construction, and operation of water supply systems are failures of the individual elements of the drinking water supply systems and, above all, threats to and interventions in the systems from the outside. In particular, after the events of the 11th September 2001 in the USA, issues related to estimating, assessing and managing of risks associated with drinking water supplies to the inhabitants have started to be formulated and addressed mainly in large municipal agglomerations with extensive water supply systems.

It is important to divide the risks into quantitative and qualitative. Quantitative risks in the process of drinking water supply are mainly represented by the lack of water and supply interruption. Qualitative risks are mainly represented by poor drinking water quality, the determination of which requires implementation of a risk analysis method addressing the drinking water production and keeping the water quality under continuous control.

2. Risk analysis

No common definition of risk exists, but for instance IEC 60300-3-9 defines risk as a "combination of the frequency, or probability, of occurrence and the consequence of a hazardous event". For the purposes of the WDS risk analysis, we have accepted this definition of risk and have expressed it as follows:

$$R=P \times C \qquad (1)$$

where R is the risk, P is the probability of occurrence of an undesired event (hazard) and C represents the consequences of the event.

The risk analysis is a structured process identifying both the probability of occurrence of an undesired event (hazard), and the extent of adverse consequences arising from the event, and then tries to answer the following three principal questions:

What can go wrong?	(Hazard identification)
How likely is it?	(Frequency analysis)
What are the consequences?	(Consequence analysis).

Risk identification and estimation are not very common in water treatment. One of the reasons is the traditionally low risk of a failure of the system treating and supplying water. However, there has been recently a general pressure exercised by the industry on cost reduction while keeping or improving the reliability, safety or efficiency of the operated system. One of the technical disciplines which may help achieve this target is the risk management. Generally, risk management is defined (IEC) as a "systematic application of management policies, procedures and practices to the tasks of analysing, evaluating and controlling risk", as shown in Fig.1.

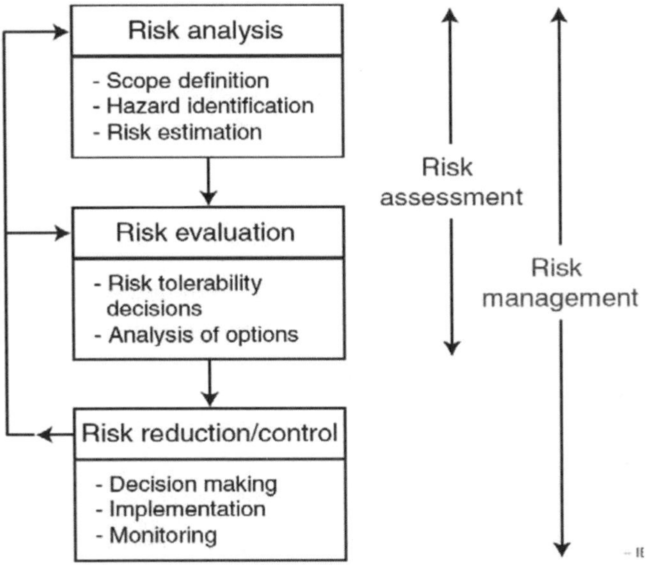

Figure 1. Risk assessment and management (IEC60300-3-9)

In Figure 1 we have illustrated the relationships between the risk analysis, risk evaluation, risk assessment, risk reduction/control and risk management according to the definitions used in IEC60300-3-9. Figure 2 shows basic steps in risk analysis and assessment procedure (Vatn, 2004).

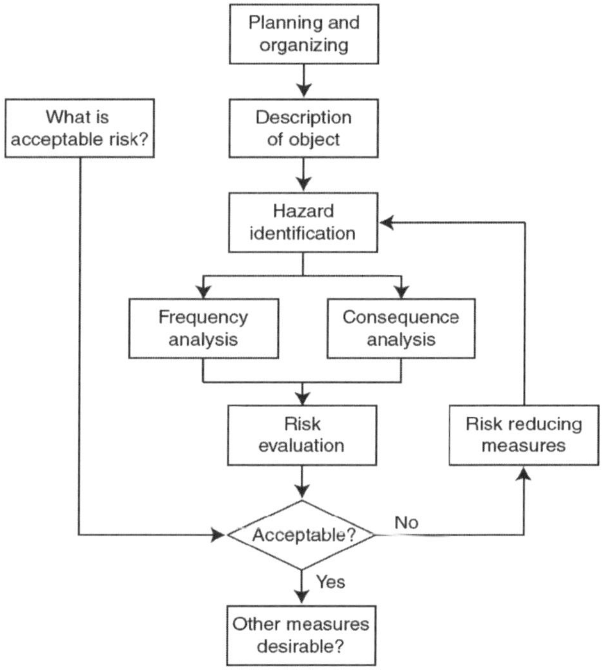

Figure 2. Risk Analysis and Assessment Procedure

3. Risk analysis techniques

There are a number of procedures for conducting risk analyses of technological systems that are suitable for identification of hazards and estimation of risks along with criteria of their selection. However, the general rule should be that a suitable technique should:

- Be scientifically justifiable and adequate for the relevant system
- Provide results in a form improving the understanding of the nature of the risk and the method of its regulation
- Be appropriate for use by other professionals so as to be observable, repeatable and verifiable.

Table 1. Review of the most frequently used techniques of water supply systems risk analysis

Event tree analysis (ETA)	A logical diagram which displays possible event sequences following a specified critical event in a system.
Failure Modes Effects and Analysis (FMEA)	Technique for basic hazard identification and analysis of frequency, analysing all kinds of failure modes of the relevant part of the system with respect to their effects on other parts as well as the system.
Fault tree analysis (FTA)	A logic diagram that displays the relationship between a potential critical event (accident) in a system and the reasons for this event.
Hazard and operability analysis (HAZOP)	Basic technique of hazard identification, systematically evaluating every part of the system in order to show how deviations from the intended proposal may occur and whether such deviations may cause problems.
Human reliability analysis (HRA)	Technique of frequency analysis dealing with the human effect on the performance of the system, evaluating the effect of human errors and mistakes on the reliability
Preliminary hazard analysis (PHA)	Technique of identification of hazards and frequency analysis that can be applied at an early stage of design in order to identify hazards and assess their criticality.

4. Risk analysis of water supply systems

Concerning the frequency and consequences, the assessment of the risk scenarios of WDS is divided into three basic subsystems of drinking water supply systems, which are as follows:

- Water source,
- Water treatment, and
- Water distribution (storage, pumping, distribution network, etc.)

Potential risk situations are divided into two categories, qualitative and quantitative risks. Furthermore, the undesired events are divided according to the origin of the hazard into three basic categories:

- Natural disasters
- Man-made threats (intentional, unintentional), and
- Technological disasters.

The text below concentrates on the category of technological disasters.

4.1. QUANTITATIVE RISK

The quantitative risk in the process of water supply is mainly represented by the lack of drinking water. Determination of quantitative risks is primarily based on the findings of the failure theory and the reliability theory with respect to the individual elements of the drinking water supply systems, which have been quite well elaborated and provide, along with mathematical modelling techniques, a very good theoretical and knowledge base for quantitative risk estimation, as shown later in a case study. One of the very useful reliability indicators is for instance a Hydraulic Criticality Index (HCI), which expresses the importance of each hydraulic node of the DWS. HCI computation is based on a calibrated hydraulic model of the network and real water demands. HCI is computed by specialized software RelNet (CARE-W, 2003).

4.2. QUALITATIVE RISK

A qualitative risk in water-supply terminology describes insufficient quality or contamination of raw or drinking water. At every stage in the production and delivery of drinking water, hazards can potentially compromise the quality of the water. Pipe distribution systems may be less vulnerable to contamination than open surface-water catchments; however, if pipe systems become contaminated, there may be no safety barriers before consumption. The most effective way to ensure the safety of drinking water is through adoption of quality assurance schemes that ensure that water supplies are designed, operated and maintained properly to prevent contamination from occurring. A very good example of such an approach is the implementation of Hazard Analysis and Critical Control Points methodology (HACCP). World Health Organization (WHO) has recently developed a risk management tool called Water Safety Plans (WSP) built on the HACCP principles. The purpose of WSPs is to minimize risks through identification and management of vulnerable points within water supply, which allows hazards.

Water quality failures in WDS can generally be classified into the following major categories (Kleiner, 1998):

- Intrusion of contaminants into the WDS through system components;
- Re-growth of microorganisms in the WDS;
- Injured microbes and residual chemicals and their by-products from water treatment plant;
- Leaching of chemicals and corrosion products from system components into the water; and
- Permeation of organic compounds from the soil through system components into the water supplies.

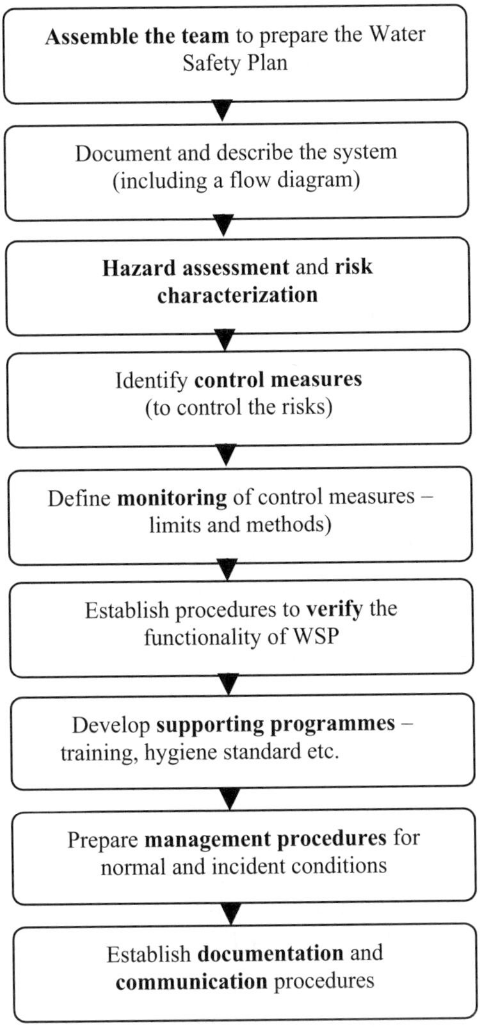

Figure 3. Overview of the water safety plan framework

To quantify the risk of water contamination in WDS is a difficult task, because of the WDS complexity and much uncertainty. An approach based on a hierarchical structure provides a breakdown of the total risk into basic risk items.

4.2.1. *HACCP Principles and water safety plans*

The HACCP approach is compulsorily applied in the Czech Republic based on the current legislation in the food industry and in sanitary requirements on food catering establishments. Most of large water utilities in the Czech Republic have implemented the HACCP in some, although partial, form. They are also forced to do so due to the amendment of Regulations of the Ministry of Health of the Czech Republic 252/2004 Coll., on drinking water, which precisely copies European Regulation No. 98/83/ES, and increase the minimum requited frequency of water quality inspections of the customers' taps. Whereas in food processing the application of the HACCP may be restricted to the production itself, the drinking water analysis must also cover risks before the production (catchments area, water source protection) and after the production (WDS, consumer's installations).

Before implementing the HACCP system it is necessary to conduct a risk analysis of a specific water system and assess potential hazards that may have a negative impact on the sanitary and hygienic quality of drinking water during production, treatment, storage, distribution and consumption. Water sources and catchments areas, raw water, chemicals for raw water treatment, and technological procedures at water treatment plants are analysed, drinking water is analysed and the distribution system and water reservoirs are analysed.

Following the hazard identification it is necessary to set up a system of critical control and protection points (CCP) in the course of abstraction, production and distribution of water, where the protection against hazards is concentrated and the set criteria are evaluated (the critical limits). These limits must be systematically monitored and they must reach the prescribed values. An integral part of the HACCP includes verification of the functionality of the system and its verification as shown in Fig. 3 (Safe Piped Water).

4.3. RISK EVALUATION

All risk scenarios are identified, clearly categorised and, based on the analysis of the frequency and consequences, they will be assessed according to Table 2, based on the FMEA principles.

Table 2. Risk identification and evaluation

	Category of risk evaluation				
Risk category	VH	H	M	L	VL
Frequency (probability)	A	B	C	D	E
Consequences	1	2	3	4	5

Remark: VH-very high; H-high; M-moderate; L-low; VL-very low

Determination of the scale of consequences is specific for every system and, to a certain extent, depends on the expert opinion and priorities of the system operator.

When identifying the critical condition of the system (hazard), a great number of scenarios of potentially undesired events may be created in practice and it may not always be easy to perform in-depth frequency and consequences analysis of each scenario.

Table 3. Risk evaluation matrix

		Consequences				
Frequency	Category	VH	H	M	L	VL
	VH	1A	2A	3A	4A	5A
	H	1B	2B	3B	4B	5B
	M	1C	2C	3C	4C	5C
	L	1D	2D	3D	4D	5D
	VL	1E	2E	3E	4E	5E

Remark: 1A – very high risk; 5E – very low risk

In such situations it is advisable to classify the failure scenarios and locate them in a risk matrix showing various risk levels (see Table 3). The output of this matrix are risks with a higher risk level, which are quantified in detail preferentially to the risks with lower risk level to be handled later or disregarded in the further steps.

What is suitable for the individually identified risks, which are included in the risk matrix, is to determine the difficulty in risk determination, both in terms of the frequency of their occurrence, and the impacts of the individual risk events. Each risk event is specific and requires a different approach to input data collection and the use of the methodology of risk identification. For this reason it is advisable to perform a kind of categorisation and to classify each event based on the difficulty of its determination into categories 1-little difficult up to 3-very difficult.

The outcome of this process is an overview table showing all theoretically feasible hazardous events in the system with their evaluation based on the risk level and difficulty of risk determination – availability of data and difficulty of the methodology, see Table 4.

Table 4. Example of possible hazardous events and their risk categories for pumping stations

Quantitative risks	Risk category	Estimation difficulty	
		Data	Methodology
Technological part			
C1-electricity outage	4D	1	1
C2-pump failure	3C	1	1
C3-engine failure	3C	1	1
C4-controlling element failure	3E	1	1
Structural part			
C5-age of building	3D	1	1
C6-man-made threats	1E	3	1
C7-natural disaster	2E	3	1

Qualitative risks	Risk category	Estimation difficulty	
		Data	Methodology
Water quality			
C8-rain water infiltration	4D	1	1
C9-groundwater infiltration	3C	1	1
C3-man-made threats	1E	3	1

Remark: 1-simple to estimate, 3-difficult to estimate

5. Case study

The case study demonstrates the method of determining the level of the probability of occurrence of hazardous events and the extent of consequences of the critical events which may occur on the supply mains of the water supply system using the FMEA method (Failure Modes Effects and Analysis). In this case study the authors concentrate mainly on the failure rate of water mains and the only evaluated hazardous event in the case study is a water-main failure (failure of the network pipe section). The basic input data for the risk analysis is the data of the technical condition of the water supply system expressed by means of several (TI) technical indicators of the network. One of the possible TIs is a hydraulic reliability of the system (HCI).

The failure rate of the water pipelines is a commonly used technical indicator, which is usually determined on the basis of statistic evaluation of the failure records. The authors present a rather different approach, where they first express the failure (hazardous event) as a function of several selected factors

"F", affecting the occurrence of failure. These are, for example, pipe age, operating pressure, bedding conditions, maintenance, etc. Each factor is allocated its weight – importance "w".

$$P = \sum_{i=1}^{n} (P_i * w_i) \qquad (2)$$

Where: P - the final category of failure occurrence probability,
P_i - the category of failure occurrence probability based on factor i,
w_i - the weight of category of failure occurrence probability based on factor i,
n - the number of factors included in the analysis of hazardous events.

Based on experience and sound knowledge of local conditions, the individual factors are classified according to specific situation (the technical condition of the pipes) and the effect on the failure occurrence into categories K1 (low failure probability) - K5 (high failure probability). The overall category of the failure occurrence probability "P" is set according to the relation (2).

The final category of the failure occurrence probability is expressed by symbols P1 - P5, in numerical terms 0 - 1. Rating a specific hazard as P1 (0-0.2) means low failure probability. On the other hand, a rating P5 (0.8-1) means high failure probability (see Table 5).

Table 5. Examples of probability of occurrence estimation – water main failures

Water main failure	**Factors – F**				
Factor designation	F1	F2	F3	F4	F5 – Fn
Factor	pipe age	pressure	bedding conditions	maintenance	other factors
Evaluating category P_i of factor i	K2	K5	K3	K2	K2
Weight w_i	0.3	0.2	0.2	0.2	0.1

The calculation of the resulting probability of failure occurrence of the supply mains, based on Table 5, would be in this case as follows:

$$P= (K2*0.3 + K5*0.2 + K3*0.2 + K2*0.2 + K2*0.1) = 0.46 \qquad (3)$$

this represents the category P3 of probability of failure.

The consequences "C" of each hazardous event are determined in a similar manner. For the needs of the case study the total consequences were broken down to four groups. They are as follows:

- G1 Personal injuries and fatalities
- G2 Socio-economic losses
- G3 Economic losses, and
- G4 Environmental losses.

Based on the specific situation, the consequences in each of the four groups G1 – G4 have been assessed and classified in categories K1 - K5 (see above). Each of these categories has been allocated a weight according to its importance. The level of overall consequences of the risk event is then set by the following relation:

$$C = \sum_{i=1}^{4} (G_i * w_i) \tag{4}$$

Where: C - the final category of consequences of undesired event,
G_i - the category of ***i*** group of consequences
w_i - the weight of the category of ***i*** group of consequences.

The total consequences of a hazardous event (failure) are expressed by symbols of categories C1 - C5, in numerical terms from 1 to 5. The rating C1 (1-1.7) means low failure consequences. On the other hand, the rating C5 (4.3-5) means high failure consequences.

The final risk assessment expresses the risk assessment matrix (Table 6), where the risk of each hazardous event is expressed as a combination of likelihood of occurrence and consequences of the event based on relation (1). The lowest risk is represented by a combination of P1 x C1 (see the left upper corner square) and the highest risk is represented by a combination of P5 x C5 (the right bottom corner square).

Table 6. Risk assessment matrix

		Consequences (category)				
		C1	C2	C3	C4	C5
Probability frequency (category)	P1	very low				
	P2		low			
	P3			medium		
	P4					
	P5					very high

The case study demonstrates an approach according to which the risk analysis of a part of the water supply system is performed on the basis of an expert estimate of partial technical indicators and their systematic evaluation. This methodology has been used for testing a concrete water supply network of a water utility in the Czech Republic and represents a very promising and flexible tool for handling the issues of risk analysis.

6. Summary and conclusions

The risk analysis is a structured process dealing with identification and evaluation of risks and consequent measures. It is generally not possible to set guidelines that could be used for risk analyses of water supply systems as each water supply system is specific, including its operating conditions. However, it is possible to develop "instructions" on how to approach the risk analysis of the water supply systems.

In the field of water supplies in the Czech Republic the performance of risk analysis is not very common yet, but in many other branches of industry it has become the established common practice. This mainly concerns aircraft industry, food processing industry, car industry, chemical industry and power engineering, and the risk analysis methods developed for these branches of industry can be successfully applied in the field of water supply.

References

Cabrera, E., Vela, A., F.: *Improving Efficiency and Reliability in Water Distribution Systems*, Dortrecht: Kluwer Academic Publishers, 1995, ISBN 0-7923-3536-8.

CARE-W: WP2 – *Description and validation of Technical tools D4* – Report on the tests and validation of technical, EU project under the 5th Framework Program, Cemagref, October 2003.

ČSN IEC 300, Management spolehlivosti - Analýza rizika technologických systémů (in Czech), Praha, Český normalizační institut, 1996.

ČSN IEC 812, Metoda analýzy spolehlivosti systému – Postup analýzy způsobů a důsledků poruch (FMEA)(in Czech).

Egerton, A., J.: Risk Assessment Techniques in the Water Industry, Swindon: Water Research Centre, 1999.

Godfrey, S.; Howard, G.; Niwagaba, S.; *Improving risk assessment and management in urban water supplies*, 28th WEDC Conference proceedings, Kokata, 2002.

Kleiner, Y., *Risk factors in water distribution systems*, British Columbia Water and Waste Water Association 26th Annual Conference, Whistler, B.C., Canada, 1998.

Sadiq, R.; Kleiner, Y.; Rajani, B., *Water quality failure in distribution networks: a framework for an aggregative risk analysis*, AWWA Annual conference proceedings, Anaheim, California, 2003, pp.1-11.

Savic, D.: *A Methodology for the Design of Water Distribution Systems under Condition of Risk and Uncertainty*, international workshop COST C19 "Proactive crisis management of urban infrastructure - Risk and reliability modelling of infrastructure and urban system", Brescia 2005.

Sedlák, M.: *Risk analysis of water distribution networks*, Diploma thesis, Brno University of technology, June 2004, Brno, Czech Republic.